TALES FROM CONCRETE JUNGLES

I dedicate this book to Donia and Nicole Lindo. I will always love you both.

TALES FROM CONCRETE JUNGLES

Urban birding around the world

David Lindo

B L O O M S B U R Y

LONDON · NEW DELHI · NEW YORK · SYDNEY

Bloomsbury Natural History
An imprint of Bloomsbury Publishing Plc

50 Bedford Square   1385 Broadway
London   New York
WC1B 3DP   NY 10018
UK   USA

www.bloomsbury.com

First published 2015

A catalogue record for this book is available from the British Library.

Library of Congress Cataloguing-in-Publication data has been applied for.

ISBN (hardback) 978-1-4729-1837-6
ISBN (trade paperback) 978-1-4729-1838-3
ISBN (ebook) 978-1-4729-1858-1

2 4 6 8 10 9 7 5 3 1

Typeset by Deanta Global Publishing Services, Chennai, India

Printed and bound in Great Britain by CPI Group (UK) Ltd, Croydon CR0 4YY

Contents

0545

Introduction

In August 2006 my life changed in a way I could never have imagined. I had just made my debut as the Urban Birder, having appeared on BBC's *Springwatch*, and was feeling totally on top of the world. I was strolling around the British Birdwatching Fair, the biggest and best birding convention of its kind in the world. Cock-of-the-rock. Visiting the bird fair was already etched into my psyche; it was the point of the year that everything and everybody remotely birdy planned for. But this year was different. I had just been on telly. Would anybody recognise me? I was especially hoping to be spotted by a scouting BBC executive browsing in the art marquee or forcing down a hot dog at the food tent as he cast a roving eye, ever on the lookout for a rough diamond. The sad answer was a resounding no to every hope I harboured. No one really recognised me apart from someone who had lent me a tenner the previous bird fair. He was clearly missing its weight in his wallet. And there certainly weren't any telly executives lurking. No matter, I still felt cocky. So when I decided to walk onto the *Bird Watching* magazine stand to cheekily advise Kevin Wilmott, the then editor, to write a piece about me, my bravado dial was set to maximum. That sudden surge of courage during my triumphant entrance immediately evaporated the moment the last of those words left my mouth. Kevin looked at me without saying a word. His face said it all: 'Who is this nutter?' Then he uttered: 'Who are you?'

Why should he have written about me? OK, I had been on *Springwatch* for five minutes but that hardly meant that I was the next Bill Oddie. He broke the quizzical stare that seemed to last for an eon, by laughing. To his everlasting credit not only did he feature me in a Q&A, but he eventually handed me my first regular writing opportunity – an urban birding column in the magazine. A brave move considering that I hadn't written so much as a sentence for public consumption before. The column initially featured my musings on the general act of urban birding but I soon thought it would be interesting to visit cities worldwide in pursuit of fresh material. Thinking that this would be a subject that perhaps would last for a year at a push before running dry, I was surprised by the striking variations between the cities I visited. Little did I know that this would be the start of a road trip like no other. Each urban centre seemed to have its own personality that was further heightened by the people I met and the luck that I had (and often didn't have) with the birds during the period of my visit. This book is a collection of many of those adventures spawned from that fateful meeting on the *Bird Watching* magazine stand. It spans from my early days with the magazine back in 2006 through to 2013, the time of writing.

Don't expect to find a compendium of birding sites within an everlasting list of the world's cities. That's not going to happen. Instead, I would like to invite you on a journey that first examines some of the principles of everyday urban birding before launching into a meander through a few of the world's cities, starting with some British ones. Not all the places I visit are even cities. Some are islands, others are specific sites that I thought rude not to pop into for a quick gander. I am not confined to cities as I feel that the city perspective on birding can also be brought out into the hinterlands. These stories are largely extended versions of the ones that originally appeared in my regular column in *Bird Watching* magazine. That said, there are also a few that have not been published before.

As I said before, this has been a journey that got started quite by accident and it is one that has brought me in touch not only with some wondrous wildlife but with some amazing conservationists. Many of these people truly impressed me with their commitment to urban wildlife conservation no matter how small their projects were. It is their work and successes that I wish to celebrate in the pages of this book. I hope that they inspire you to look at cities with different eyes and realise that the conservation message is perhaps more important to spread here than anywhere else in the world.

Being an Urban Birder

When was the last time that you stood in the middle of a busy street in a busy urban centre with your eyes closed listening out for the songs and calls of hitherto invisible birds? Let me rephrase the question: when was the first time you ever seriously indulged in urban birding?

If the answer is never then that's fine because urban birding is a bit of a mindset thing. Let me explain. I was born in London and spent my formative years watching birds in urban environments. Although I grew up to eventually call myself the Urban Birder, in reality it took me years to truly believe that I could find birds in cities. That was borne from the fact that I had no mentors and very little reference points as to where I could find birds in the metropolis. While wandering around in my daily city life I noticed birds. I had unconsciously discovered that birds were indeed everywhere.

The day I woke up to the idea of birds in the city was the day I fell in love with Wormwood Scrubs, my local patch. This area of land completely surrounded by housing, industry, roads, a hospital and a forbidding prison of the same name still managed to attract birdlife that even my friends in Norfolk were jealous of. When you start to see the urban world as a habitat with cliffs, woodland, marshes, lakes, rivers and scrubland, that is when you start to see birds.

Uncle Birds

I've been a birder since the devil was a boy, which comes as a surprise to those who only know me from behind the wheels of steel when I'm spinning tunes in a nightclub or diving around between the sticks on a football pitch on a cold, wet Saturday morning.

If only they knew that early every morning I am to be found roaming my beloved local patch, Wormwood Scrubs in west London; home to the notorious prison of the same name, mentioned in The Jam's 'Down in the Tube Station at Midnight' and featured in the original *The Italian Job* starring the King of the Cockneys, Michael Caine.

Sometimes after visiting the Scrubs, having seen next to nothing, I ask myself: 'Why am I here birding in the middle of London when I could be watching real and more exotic birds somewhere decent out in the countryside?'

To answer that question, I have to go back to when I was a little five year old staring out of my bedroom window over my back garden in suburban Wembley, north London. At first I was spotting 'mummy birds' and 'daddy birds' (Starlings and Blackbirds respectively), 'baby birds' and 'uncle birds' (House Sparrows and Carrion Crows). I watched Woodpigeons performing their display flights and christened them 'jack-in-the-boxes', and the rounded wings of the passage Lapwings that I occasionally saw reminded me of spoons, so they were quickly renamed 'spoon wings'. A whole new world was opening up before me.

Eventually, when I got hold of an old field guide there was no stopping me. By the age of eight I had begun keeping a list of the birds I recognised in my garden – including a few really dodgy sightings.

I quickly learnt that the best time to observe the action was first thing in the morning before Mrs Smith next door came out to mow her lawn. The drawback with pre-school early-morning birding from my bedroom window was the potential for being identified as a peeping Tom.

By the time I was ten, I had progressed to my local park, consisting of mown grassland, a river and large areas of undeveloped wasteland. In those days I took for granted the breeding Skylarks and wintering flocks of Tree Sparrows. It was here that I discovered that migrants like Yellow Wagtails and Wheatears were more regular than I had ever imagined.

However, when I had reached thirteen, the Skylarks and Tree Sparrows had disappeared, the wasteland had been built upon and the new tide of human residents had invaded the park. The appeal of my nursery local patch had diminished. The major turning point in my early urban birding occurred after I read *Birds of Town and Suburb* by the ornithologist and broadcaster Eric Simms. I learnt three important lessons.

First, look at every bird you see – even if you think you know what it is. Anything can turn up anywhere at anytime and at the very least, you may learn something new about a species that you thought you knew well. Second, you can watch birds anywhere, even within a large city and, finally, he taught me the value of patch watching.

By this time I was truly an Urban Birder and despite regular trips to Norfolk, Kent, the Isles of Scilly and the like, I still watched birds predominately in London. There is always something special about seeing birds we are more accustomed to seeing in wild environments against the backdrop of urbanisation.

I remember racing back from Norfolk after watching Marsh Harriers and Bearded Tits to see a grubby female Garganey that had decided to temporarily reside among the semi-submerged abandoned shopping trolleys and dumped scooters at Brent Reservoir.

Well, fifteen years, one Ortolan Bunting, a low-flying Honey-buzzard, and quite a few Redstarts, Wheatears and Ring Ouzels later, it's fairly clear that my decision to watch the Scrubs was a good one. But I believe that I could have picked any underwatched urban area with sufficient habitat and still have found interesting species.

So my message is this: birding is exciting and rewarding. By getting to know a local patch you will become familiar with the commoner species and you will be amazed when you witness their habits and the seasonal changes in their populations. With this ever-growing knowledge will come a great sense of pride when you eventually uncover an unusual bird.

Love Story

It's the light. There's something about the autumnal light. It sends out a vibe that is so inviting, alluring, even seductive. There's also an air of expectancy. The smells differ, as does the atmosphere and, of course, the birds; they are on the move. Something stirs within me too. I begin to develop an inexplicable excitement that begins as a twinge, a slight urge that at its height becomes an unrestrained desire to be out birding at the crack of dawn every morning regardless of the weather and despite the fact that I may have gone to bed in the wee hours the night before.

This psychological change in me begins in mid-July, when I start craning my neck skywards. Even a Woodpigeon flapping overhead is enough to have me temporarily searching the skies for hitherto undetected migrants. This peculiar behaviour has its roots in my childhood. As a boy, I noticed from my north-west London bedroom window that from June to late July there were a lot more feral pigeons flying around than at other times. At the time I put it down to migration and not to the fact that there were just a lot of pigeons flying around. But it was while watching them that I first noticed real migrants. Nowadays, by late August I become very restless. If I'm in a building I often find myself drawn to the windows where I stare out and drift away, dreaming of being somewhere looking for migrants while searching the skies for, well, anything animate with wings. My lunchtimes could easily be spent on an office-balcony vantage point surveying the urban skyline.

Autumn is definitely my favourite time of the year, as it is the season of love. I have fallen in love during this magical period on many occasions, and when I do I don't hold back, I just get stuck right in. Oh, let me explain what I mean by love. It is unconditional: it can blind you, lead you astray, distract you and in extreme cases cause you to defend the object of your desire to your last breath. Allow me to tell you a love story.

This story concerns the conducting of a fingertip search for migrants in one of my favourite places on the planet: Cape Clear, that amazing gem of an island off the coast of County Cork, Ireland. I fell in love with this place after reading for years about the birds that have been found there. I decided to experience it for myself for a week in one October. Well, my original plan was to spend a week in County Cork, incorporating a couple of days on the Cape. However, by the time I got to Baltimore to catch the ferry across to the island, the lure of potential island vagrants was too strong, so at the last minute I decided to spend my entire time there.

When I stepped off the ferry at the other end I was immediately struck by the friendliness of the locals. As I walked to my B&B I passed by walled coastal gardens that were filled with feeding Siskins, and above my head swirled Herring Gulls, Jackdaws, Choughs and a few Ravens. Walking the incline that led to my accommodation, I noticed some very interesting looking gardens filled with the kinds of bushes that migrants love getting lost in. A cursory glance resulted in foraging Robins and Blackbirds. Yet even these *ordinary* birds enchanted me, filling me with delight as they fed in the autumn light. I was blown away and I hadn't even started birding properly yet.

Over the ensuing days, I walked the length and breadth of the island many times, enjoying birds as diverse as Little Grebes and Fieldfares through to Merlin, a Short-eared Owl, Yellow-browed Warblers and a lonesome Pied Flycatcher. Even dipping out on a male Ring Ouzel and a juvenile Night Heron didn't dull my enthusiasm for my newly found nirvana.

One day, while walking past a farmhouse, I was called over by a distressed looking elderly local. With telescope draped across my shoulder, I went to her aid. After introducing herself as Mary (confusingly, most women on the island were called Mary), she explained that her husband had fallen out of his bed and was too heavy to lift up. Undaunted, I followed her into the house. Her old man, who must have been in his nineties, weighed more than a small ox and could I budge him? In the end I had to enlist the help of Mary's daughter (another Mary) and her husband. After shifting her husband back onto his bed, I was invited to stay for tea by a very grateful Mary senior.

Every evening after a thoroughly enjoyable day in the field, I would turn up at the Bird Observatory for the log call with the other visiting birders. The evenings would invariably descend into a garrulous and raucous get-together hosted by warden Steve Wing and partner Mary (!). Before long, my week had ended and it was time to head back to London. As I went for my final walk around my by now Irish local patch, I could not help but feel a great sense of connection and affection for the place and, despite not finding the vagrants I dreamt of, a piece of my heart was embedded into the fibre of the island, forever present in the glow of the autumn light.

Birds on Film

I'm seven years old and cowering behind the sofa at home. Moments earlier, I was sitting on the sofa watching television. Why am I now behind the sofa? There was the clearly audible sound of thousands of beating wings, deranged squawking and people screaming. I'm peering over the top of the sofa to see hundreds of gulls and crows wantonly attacking the hapless panicking humans, pecking and scratching with vengeance. I am watching Alfred Hitchcock's immortal *The Birds*.

As is the case with countless others, my irrational nervousness towards anything with wings that chooses to flap around my head was born right there. For a small kid like I was then, that fear was magnified because the reasons for the attacks were unresolved in the film. Years later, I watched *The Birds* again and admired Hitchcock's mastery of suspense and smiled at some of the obviously stuffed birds swooping around without even once covering my eyes. I have since largely got over my phobia of flying things; however, even to this day the sound of multitudes of flapping wings still leaves me feeling slightly uneasy.

Birds have often been portrayed in film either as malevolent creatures or as avian eye candy frequently chucked in with little regard to whether they are relevant to the story or not. Much to the annoyance of the birders among us, some film-makers blatantly get it wrong, featuring birds and other wildlife that clearly don't belong. I mean, why have some boring brown bird when you could cast a gaudy interesting species that has 'camera presence'. Besides, how many of the cinema-going public would ever have noticed anyway?

One of the most obvious cases of total miscasting was in *Tarzan and the Amazons*. Made in 1945, the film sees Tarzan supposedly frolicking in the Amazon looking for a lost city of Amazon women while encountering wildlife more befitting the African continent. In the opening scene, Cheetah, the chimp, was fishing in a river that I presume was the Amazon, in the company of some parrots. The macaws were reasonable to include but the cockatoos, which of course are natives of Australasia, grated on me. The year before saw the release of *Tawny Pipit*, directed by Bernard Miles and Charles Saunders. The plot follows the story of a Second World War fighter pilot, recovering from his injuries in a sleepy English village, who discovers a pair of rare Tawny Pipits nesting in the adjacent countryside. The featured birds were in fact Meadow Pipits.

More recently, while watching Anthony Minghella's *Cold Mountain*, which was set in North Carolina, I noticed that many of the corvids in the mountain forests were Hooded Crows that are natives of the Old World. And then there's the classic Pygmy Nuthatch scene in *Charlie's Angels*, which I find difficult to erase from my mind. Boswell is banged up in an undisclosed prison cell somewhere in North America and manages to contact Cameron Diaz's character. She precisely locates him by hearing a calling Pygmy Nuthatch recognised through a walkie-talkie. Now had she heard a Zapata Wren, found only in the marshland of the same name in Cuba, or Socotra Cormorants that are mainly found, you guessed it, on Socotra Island in the Indian Ocean, then I would say fair dos. I could have lived with the fact that the Pygmy Nuthatch has a patchy, though not a particularly restricted, range in western America. However, when the bird was revealed it was not even a nuthatch but an American icterid. I'm sure the TV armchair birders out there could fill the pages of this book with many other such moments.

Stuffed, mechanical and animated birds in films seem to be commonplace too. Although usually live, the caged lovebirds featured in *The Birds* became obviously stuffed and swayed in unison in a speeding car. In David Lynch's *Blue*

Velvet, the warbling electronic American Robins were amusing to watch, but don't get me started on the owls featured in some films that are often nothing short of laughable, though the Eagle Owl in Ridley Scott's superb *Blade Runner* was both real and absolutely stunning.

There was also a wealth of birds to be watched in Chris Weitz's *The Golden Compass*. Seeing as this is a fantasy film, there were plenty of dreamt-up birds, including dozens of weird passerines, odd-looking Red Kites and a first-winter gull with a distinctive head pattern that was not dissimilar to a Dotterel. Oh, and if you ever get to see *I Am Legend*, starring Will Smith, check out the array of birds that are coincidentally featured, including the continuous sound of a daytime churring nightjar of some sort.

One thing that I will give good old Mr Hitchcock is the fact that in the main, he used species that were relevant to the area in which *The Birds* was set – coastal California. I loved the fact that he featured Western Gulls, the typical west-coast larid. Gus Van Sant's *Finding Forrester*, released in 2001, featured a scene in which the lead character, played by none other than Sean Connery, pointed out a proper singing male Yellow Warbler to an urban kid from his New York office window. In Tony Scott's *Enemy of the State*, at least the Canada Geese were used in the correct settings and, what's more, one of the central characters was a birder. Unfortunately, he met a very grizzly end fairly early on in the film. But for me, Ken Loach's classic *Kes* stands out. It felt totally real and made me look at Kestrels in a different light.

So I suppose the message to all you film-makers out there is simple: get it right guys to stop us birders from bursting out laughing in packed cinemas at inappropriate moments. I hear that Hollywood is thinking about remaking *The Birds*. When that hits the big screens will there be a whole new generation of people with a fear of flapping wings?

Viva Ronaldo

It's 1.30 a.m. on a dank, dismal May night, but frankly I couldn't care less. I'm euphoric. 'Viva Ronaldo' the crowd chants in unison as I wildly embrace similarly excited and jubilant complete strangers. I'm alternating between sitting and standing in the Luzhniki Stadium in Moscow having just watched my favourite football team, Manchester United, beat Chelsea to lift the European Cup. Don't worry, you haven't picked up a sports book by mistake and I'm not going to start waxing lyrical about pass completion rates, noteworthy goals and other soccer-related trivia. No, this is a tale about finding happiness.

My three-day Moscow trip came about quite by accident, as I hadn't planned to go to the Cup Final until a ticket suddenly came into my possession literally a couple days before the game. Being a United follower was one thing, but being an even bigger urban birding fan meant that I could not turn down the opportunity to discover the birdlife in this historic city.

As I boarded the plane dreaming of my team lifting the Cup, I drooled at the thought of seeing several of the eastern European woodpecker species that had so far evaded me, and bumping into Bluethroats and perhaps spying on Red-footed Falcons as they frolicked over some parkland heath. To be honest, I didn't have a clue what to expect. I was just excited about the prospect of visiting a Russian city.

I landed hours before the game in an unpronounceable Moscow airport on the outskirts of town to be greeted by stony-faced officials and less-than-happy-looking police officers. This initial impression of the Muscovites was to be a reoccurring theme. While walking to the taxi rank outside the airport a hidden Siskin singing from a nearby block of conifers momentarily lifted my heart. I felt that I had encountered at least one friendly-sounding voice.

Due to time differences (Moscow is three hours ahead of London) the match started late in the evening and I eventually arrived back at my hotel in the east of the city at 3.30 a.m. after being herded out of the stadium with the other supporters like a herd of Friesians. As dawn was just two hours away, I decided to cap the celebrations by popping out to a local, seemingly nameless wooded park that I noticed on the hotel's tourist map. So, I duly strolled out of the hotel at daybreak and headed down the main road.

Now I use the term 'daybreak' loosely as it was still dark, with impending rain clouds ominously hanging around overhead. Despite the early hour, the main road was already filled with buses, trams and military vehicles, as well as chugging diesel cars, all slowly crawling past block after block of pre-Iron Curtain housing estates. Their drab box shapes were occasionally brightened by a cawing Hooded Crow and a few skirting Swifts.

The most notable thing, though, was the genuinely unhappy-looking commuters who didn't have a smile for anyone and seemed to have the entire weight of the world on their shoulders. Given Russia's history, what with hundreds of years of tyrant rule, slaughter and invasions, plus its particularly long, dark and nippy winters, I can totally forgive them for being miserable. Eventually, I reached my adopted patch. Numerous House Sparrows and feral pigeons greeted me as I walked down the concreted parade leading to the large and perhaps once ornate gates of the park. I could see why this place wasn't on the tourist map – it was a *real* inner city 'local' park replete with drunks, shell-suited, balding, bland-faced joggers and grim-looking dog walkers.

A Chaffinch burst into song from the first couple of trees I came to. Its voice was distinctly different from the voices of the birds that I knew back home. Dare I say, harsher sounding. A White Wagtail or two cavorted on the paved road that seemed to dissect the area. These were the only birds I saw bar the numerous feral pigeons. Yes, the initial birding was not promising. But the urban birding mantra states 'anything can turn up anywhere', so I ventured further into the park. Soon the terrain began to change and before long I was in deeper woodland and the outlook had changed. First a Nuthatch appeared, then I flushed a Fieldfare from the woodland floor. Tree Sparrows, Great Spotted Woodpeckers, a pair of Pied Flycatchers and singing Willow Warblers also quickly revealed themselves. It almost felt as though I had gone through an invisible door and had entered into a secret world.

Suddenly, an explosive 'chok, chok, chok' filled the air followed by an equally loud and repetitive series of notes that could only have been uttered by a Thrush Nightingale. Within ten minutes I was watching my first proper Sprosser – not counting the sorry individual under a bush at Salthouse, Norfolk, that I twitched in 1992. As I stood watching this Russian bird, mesmerised by its incredible song, its barely rufous tail trembled and its mottled breast swelled with every sweet note. It quite literally filled me with joy; how could the Muscovites be so unhappy when right here in the middle of their city they could fill their hearts with glee just by listening to the birdsong? No sooner had I thought that than the rain began to pour. A sudden and massive deluge of water. I was totally drenched as I scurried back to the hotel.

The remainder of my time in the Russian capital was punctuated by almost continuous rain, ranging from drizzle to torrential downpour. Despite the weather, I religiously covered my patch, getting up close and personal with the rather plentiful nesting Fieldfares. I am used to seeing these handsome thrushes in roving winter flocks in the UK, so it was a bit of a shock to the system to see them on home territory, 'warts an' all'. The Spanish call them Royal Thrushes, yet here I was watching their domestics: manky-looking adults colonially sitting on nests, singing their grating song and feeding scruffy fledglings.

Aside from finding an Icterine Warbler and hearing a glorious male Golden Oriole's fluty call, my overriding Moscow memory was the singing Thrush Nightingales (and a European Cup). My dilapidated wooded patch was ample proof that you don't have to be anywhere special to appreciate birds.

Catch the Pigeon

Do you ever get the feeling that you are drawn to things for inexplicable reasons?

For several months I had increasingly begun to think about feral pigeons. I had been taking more notice of them as they flew over my local patch; I even put my newspaper down one morning while waiting in an underground station to observe a ropy-looking pigeon tackle some discarded cheese and onion crisps strewn around the heels of unconcerned commuters. When it approached me I didn't even have the urge to shoo it away.

What was happening to me? I hadn't registered feral pigeons for years. I never count them on my bird lists and they were lucky to get a cursory glance in the field. I mean, if you were to take to the streets with survey clipboard in hand to ask members of the public what they thought was the most disliked bird in Britain a frequent answer would be the flying rats – those dirty street pigeons. Some respondents would probably summon up depressing images of piles of soiled nest material under bridges, long-dead birds blocking gutters and the end credits of *Coronation Street*. I have to put up my hand and say that I would have also been one of the lynch mob cursing the ground they excreted on.

On a recent shopping expedition I decided to nip into a bookstore to break up the monotony. I headed for the natural history section – naturally – and picked up the first bird book I saw. It was a guide to the birds of Britain. When I randomly opened it I was treated to a plate featuring pigeons. My eye was drawn to the feral pigeon text and I was amazed to read that there were supposedly 3.5 million pairs in Britain. How could that be? I hardly see pigeons these days. A couple of days later I happened to pick up a national newspaper and there on page two was a picture of a feral pigeon above the caption 'Woodpigeons outnumber their feral cousins by two to one in urban areas.' Intrigued, I gave my considered friends at the British Trust for Ornithology (BTO) a call and learnt that at most, there are 250,000 pairs of feral pigeons in Britain. After years of assuming that there were 250,000 living along my street to discover that they are not as common as I once suspected was a bit of an eye-opener.

Well, I decided to take a second look at this most familiar of birds in the UK, if not the world. Although to some they are the embodiment of all that's dirty and grimy, there is a lot to be said for our humble pigeons. They are wartime heroes, for God's sake. And after the chicken, they have the distinction of being the first bird to be domesticated by man. They come in all different hues, provide hours of fun for pigeon fanciers, and if they are fortunate enough to have white plumage they are revered as messengers of love and peace.

They are bright and have been found to be able to recognise the human faces that feed them in a crowd. They have even allegedly learnt to use the underground system in London by deliberately getting on trains and getting off at specific stops. OK, they can be irritating as they coo incessantly outside your bedroom window on a Sunday morning when you are trying to sleep off the heavy night before, and their droppings do cause damage to buildings. And let's not forget the germs, though research has shown that this aspect of the birds is not quite as bad as people first suspected. So is it their fault? Can they be excused for doing what comes naturally? Or are they the victims of our excess?

I guess that you're getting my drift now. My views on feral pigeons have softened and I don't see them as vermin anymore. I admit that I have found my peace with pigeons. I'm learning to enjoy them again and marvel at their superior aerial ability. They are amazing flyers and that's not just when they are being chased by Peregrines.

Dangerous Urban Birding

Recently a new query was added to the usual repertoire of questions I get asked about city ornithology: is urban birding dangerous?

It was a question that stopped me in my tracks and one I seriously had to ponder. A quick survey among my birding friends soon resulted in various tales of woe, most of which were quite humorous, such as being sprayed with pig slurry and being shouted at by angry residents who were unfortunate enough to have a rarity in their neighbourhood. There was even a story of a birder being publicly assaulted at a twitch by his long-suffering wife.

If you have your wits about you, avoid certain areas after dark and don't flaunt flashy optical and photographic equipment as an invitation to be mugged, then you should be totally fine. Being barked at by German Shepherds, dive-bombed by nesting urban gulls and nearly hit by the occasional stray football is usually as good as it gets.

However, the world was a different place when I was a keen ten year old. I delighted in watching the sparrows in my backyard, studying their plumage details, but I also yearned to head further afield. I was especially fired up after reading books that spoke of people seeing, to my young mind, exotics like Cuckoos and Little Ringed Plovers. So, I used to tell my mum that I was cycling around the block to go birding at our local park with my Irish schoolmate, Alan, and instead we'd travel across London to arrive at Rainham Marshes in Essex two hours, three buses and some twenty miles later.

In those days, when visiting this pre-RSPB feral wilderness you had to make a choice: enjoy some potentially great birds while warily looking over your shoulder or turn around and go back home. Rainham was a lawless derelict land frequented by air-rifle-toting Essex boys. One day after catching sight of our first ever Cuckoo, Alan and I had to beat a hasty retreat after coming under fire from a bunch of hostile local lads. We were literally chased out of town.

Thereafter, Alan and I took the unprecedented step of 'arming' ourselves against any future attacks while out birding. A trip to the local outdoor shop ensued and we soon walked out the proud owners of a small penknife and a catapult each. I guess our purchases gave us a sense of security, though I would not condone buying that stuff today. Quite what use a catapult would have been baffled me even then – it was Alan's idea.

Thankfully, we never needed to use our weapons, which were tucked discreetly in our birding jackets, apart from once several years later when we used the knives to cut a drowning Long-tailed Duck free from a fishing net off the shore at Dunwich, Suffolk.

In fact, it was out in the countryside that I had my next brush with potential danger. Not too long ago while driving to a twitch in deepest Norfolk, my companion and I took the wrong turn and accidentally ended up on a driveway to a property hidden deep in a wood. Some tethered dogs started to bark and snarl ominously as we approached, and a guy appeared, shouting and calling us 'twitters'. He then unreasonably threatened to shoot us before running back into his house. Naturally, we didn't stick around and opted instead for some nifty rally driving in the opposite direction.

You really do need to be careful when you are abroad in a strange city where you could inadvertently wander into dangerous situations without realising it. The general rule is to either do your birding with a local, or stick with the recognised areas like nature reserves or in safe areas like the universal botanical garden that is found in almost every major city.

When I'm birding in a foreign city I always err on the side of caution to try and blend in with the locals. I came unstuck, though, a couple of years ago. One morning while on a trip to Bratislava I decided to venture into the wooded wasteland adjoining my patch, which extended for a couple of miles to a motorway flyover. While exploring, I came across a fairly recently deserted campsite that obviously had previously been inhabited by junkies. Feeling uncomfortable, I decided to head back out of the area.

Soon, as I was walking along a track I heard the sound of a Land Rover behind me. I felt nervous, as there was something very wrong about the whole situation. There were two burly-looking guys in the vehicle and neither looked particularly pleased to see me. The vehicle accelerated and eventually screeched to a halt alongside me. One of the guys jumped out and aggressively addressed me in Slovakian. There I was, in the middle of a wood in eastern Europe, with no ID, unable to speak the lingo and potentially about to be kidnapped or murdered.

I pleaded my Englishness and the fact that I was only a birder, but that resulted in not even a flicker of compassion on the face of my would-be murderer. I frantically began to think about my next move until I noticed the word Policía printed in small letters on his T-shirt. They were two undercover policemen. Somehow I didn't feel any safer. In an aggressive broken English, he told me to leave the wood and not to come out again without a passport. The lesson? As quoted in the film *An American Werewolf in London* – don't stray off the paths.

Identifying Birders

Some birds defy description, and the same can be said about some of the people who watch them. Broadly speaking, there are some people who look like classic birders. It's a look that's very difficult to articulate – they just look like birders. And there are others who despite what they tell you, simply don't fit in with the conventional stereotypes. I guess that for a lot of people I didn't fit into any convenient pigeonhole either, especially when I was younger. I fondly remember the days when I was head of membership at the BTO, travelling around Britain giving talks on the work of the Trust. Some of the people I met during my travels visibly raised their eyebrows when they clapped eyes on me. They did not expect to see a funky black dude wearing a pinstripe suit with short, dyed blond hair (yes, I had hair and used to be blond) with a pair of bins around his neck chatting about the decline of the Skylark.

I too am very guilty of stereotyping. While filming a piece for BBC1's *The One Show* the director told me that I'd be meeting and interviewing a doctor of ornithology. When I met the good doctor I was surprised to be greeted by a cool young guy in his late twenties with shoulder-length hair and several piercings in both ears and left nostril – a far cry from the heavily bearded venerable gentleman that I had expected.

I'm sure that you have all experienced situations where a non-birding acquaintance has said: 'Hey, you have to meet such and such, he's really into birds like you…' Invariably, when you do get to meet 'such and such' they turn out to be what you expected, either as fanatical as you are (which is rare), or more likely someone who in their dim and distant past may have mentioned that they liked keeping budgies.

One Saturday morning, while I was knocking back a hearty breakfast with the lads after playing football, I got chatting with a female friend of mine who at the end of the conversation dropped in those immortal words: 'Oh, you must meet my friend Jane, she's really into birds…' Smiling, I wearily asked her to get Jane to give me a call sometime – then promptly forgot about it.

Several weeks later I got a call from Jane who, after exchanging pleasantries, invited me over to her house in west London the following day for tea and a birdy chat. In my mind's eye I had an image of a middle-aged, middle-class housewife who basically wanted someone to chat to about the Robins in her garden. The next day saw me parked outside her detached modern super-swanky town house. From the outside it was pretty imposing with a gated entrance and three big-deal cars in the driveway.

As I walked towards the house I was mentally checking to make sure that I had the right address. I was greeted at the door by a butler. Behind him, sporting the broadest of grins, was an attractive young woman with brown shoulder-length hair.

We went into her study, filled wall to wall with bird books, and she challenged me to correctly identify her collection of stuffed birds. I was surprised but played along. Thankfully, the first set she pulled out was relatively easy as it included superbly preserved Goldcrest, Long-tailed Tit and Jay specimens. Then she pulled out her woodpeckers. She explained that wood-peckers were her favourite birds and that she made biannual trips to eastern Europe to study them. I began to sweat as I scanned the unfamiliar European ones. The three British woodpeckers and a beautifully presented Grey-headed Wood-pecker were quickly identified and I eventually sussed the White-backed. But I failed miserably on the Middle-spotted, calling it a Syrian. She declared that I had passed the test and earned my cup of tea.

Once in her kitchen I gazed out of her wall-to-wall French windows into the garden, a plot of land that seemed like half a football pitch. The perimeter of the garden was lined with tall old trees, each furnished with woodpecker, tit and bat boxes. Working the land were a couple of uniformed urban gardeners who were planting native flora specially chosen by Jane for the sole purpose of attracting wildlife. She was having weeds deliberately planted and was planning to devote a corner of her garden for perpetuity to the cultivation of berry-bearing bushes in the hope of attracting winter thrushes and even Waxwings.

I had clearly got it wrong. From now on I will look twice, maybe even three times, before I start trying to identify birders.

The Ones that Got Away

During one spring I led a London Natural History Society bird walk at Wormwood Scrubs. I was secretly hoping that it would be cancelled due to the continuous heavy downpour that had started earlier that morning. But alas, when I got to the meeting point I was stunned to see three hardy birders waiting for me in the pouring rain. I went ahead with the guided walk and initially the only calls I heard were the overriding ones from my bed beckoning me back for more slumber.

While walking through a lightly wooded area, I flushed a medium-sized, long-tailed bird that flew through the trees away from me. It all lasted a heartbeat. For a second I was flummoxed. What was that bird? Was it just one of the many Rose-ringed Parakeets that had chosen to colonise my beloved, and until fairly recently, parrot-free patch? Or could it have been a Cuckoo – an extremely rare bird down my way with less than four records in the past fifteen years?

This is a familiar scenario known to every one of us, young and old, experienced and novice, and we are assured to have many more such scenarios throughout our birding lives. Seeing birds that you just can't name is a really frustrating thing. Seeing something interesting flying over that doesn't call or viewing a bird for a tantalising few seconds can be absolute torture.

As I was preparing to consign my sighting to the 'what the hell was that?' category the object of my conjecture suddenly shot out of the other side of the trees, flying high into the open sky, seemingly battling against the elements. As I watched it through my steamed up and wet optics, it became clear that I was beholding a bedraggled Cuckoo, my fifth at the Scrubs, continuing its journey north.

On that occasion I had the luxury of a second chance. Most of the time you never get to clap eyes on your mystery bird ever again. You are left holding your hands up in despair. When out in the wilds you expect to see things that you don't initially recognise – no matter how 'expert' you are, especially when you are in a rich migrant hotspot. I'm convinced that had I had seen my Cuckoo in exactly the same habitat in the middle of rural Yorkshire I would have expected to see it, therefore a split-second view would have triggered something in my brain that may have led me to a Cuckoo.

In an urban setting you have an added problem – being confused by seeing birds that are out of context against an urban backdrop. For some reason seeing familiar birds in unfamiliar surroundings can be quite unsettling. Examples could be an Oystercatcher striding around on an inner-city football pitch, a Marsh Harrier coasting over a trading estate or a Woodcock standing outside a London underground station. One of the weirdest circumstances of a bird being found in an incongruous place befell the finder of a Red-footed Falcon during the Seventies that had just been attacked by a cat and was cowering under a bush in a Watford garden.

I remember strolling through my patch one September morning and flushing what I thought was a Linnet from the path in front of me. It flew up and landed on a fence. I immediately realised that it didn't 'feel' like a Linnet so I checked it with my bins. What I saw caught me by total surprise. It was a bunting with a buffy wash, a yellow eye-ring and moustachial stripe. For a few long, agonising moments I took leave of my senses. I just couldn't work out what this bird was. Was it an escaped cagebird? Some weird hybrid maybe? Suddenly, logic kicked in and after regis-tering its salient plumage details it became apparent that I was looking at a bloody Ortolan Bunting.

Wait, an Ortolan Bunting? But I'm in the middle of London. Had I been on the Isles of Scilly I would have probably started with Ortolan Bunting before trying to convince myself that I wasn't looking at something vastly rarer.

I can't write about the ones that came back and the ones that eventually got identified without talking about the ones that actually got away. One cloudy September morning while watching my first Turtle Dove of the autumn head west over the Scrubs, I picked up on three dots very high up heading my way from the west. I abandoned the dove to track the strange dots as they approached. Soon, I began to see their shapes. They were like long-tailed, pointy-winged dark hirundines that flew in a progressive flap, flap, glide-like fashion. I was confused. My internal computer was going into overdrive. Could they have been a group of Hobbies? Something was odd about them as their shapes were so unusual.

Despite coming closer, they were still too far away for me to see any colour on them against the grey skies or to hear a call. Suddenly, they banked and started to head back west again. 'Where are you going?' I screamed in my head. My pleading went unanswered and my mystery birds quickly disappeared from view. An awful realisation came over me. Had I been watching my first ever Bee-eaters in London?

When I got home I pored over the Bee-eater entry in *Birds of the Western Palearctic* (BWP) for more clues. What I read stopped me in my tracks, '… when on migration, can resemble hirundines with bursts of regular wing-beats to maintain rapid progress often at great height'.

Ah well, sometimes these things are sent to try us.

My Hate Affair with Parakeets

Coming to a town near you – the green-eyed monsters.

Shakespeare may have coined that derogatory moniker first in *Othello*, but it's a fairly well-known fact that I view our latest avian alien invader in an equally dim light. No, I am not the keenest fan of *Psittacula krameri*, also known as the posh pigeon, flying secateurs, long-tailed green rat, ring-its-neck parakeet or just plain Rose-ringed Parakeet. I developed an almost instant dislike for them – a distain that I used to reserve only for feral pigeons – when I first became aware of their shrieking flocks. Soon after, I was deliberately ignoring them and openly cussing every time a calling party passed overhead.

My previously parrot-free local patch has now become a parakeet conurbation in the space of three short years. While doing a bat survey the other night, I witnessed more than 2,000 birds noisily swarming in from all points of the compass to converge on a tiny copse containing just a handful of mature trees. I was horrified, but secretly had to admit that it was an amazing spectacle. But what are these birds actually about? Should I continue to publicly denounce them, or take a more liberal view and embrace them as newcomers to Britain?

Many city dwellers who were previously oblivious to any urban wildlife, unless it was a fox or a pigeon, now know about parakeets. People have remarked on this exotic's vivid green plumage as well as its less than pleasant voice. I now laugh at the stories I've heard about the public's first encounters with this distinctive bird. The accounts invariably involved lots of cooing, admiration, utter disbelief and, on occasion, distressed phone calls to the RSPCA to report missing pets.

I first started to notice Rose-rings in the late Seventies as an extremely rare vagrant at my then local patch, Brent Reservoir in north-west London. In those days you were more likely to see escaped budgies, lorikeets and unidentified Amazon parrots. Little did I know that a small population of Rose-rings squawking in a corner of south-west London was to explode exponentially thirty years later. Of course, these unquestionably attractive birds have been the subject of many urban myths surrounding how they came to be here in the first place. Jimi Hendrix, spiteful wives, ditched aircraft fuselages and film studios with doors left open have all been bandied around as reasons for the origins of our feral populations.

The real reason is actually quite boring. The birds that we see today are descendants of escaped birds. Interestingly,

Great Yarmouth played host to the very first reported nesting Rose-rings – way back in 1855, with birds reported in London a little later. How did those birds get there? Well, they came to these shores by boat disguised as the pets of sailors, only to jump ship once in sight of Blighty.

Some people may be coming round to my way of thinking. I'm continually asked about how the birds survive here and, how they affect our native birds. There are people who are now actively complaining about the noise the birds make and the hijacking of their bird feeders. Some have even enquired about the possibility of a cull.

I think that the damage has been done and there are just too many to kill off now. The Department for Environment, Food and Rural Affairs (Defra) is currently undertaking some research into the Rose-rings' impact on our natural fauna, but there is a plethora of anecdotal evidence that suggests that the parakeets are ruling the roost. Judging from hearsay, very few of our avian predators are adding Rose-rings to their snack list. I was recently told about an opportunistic second-winter Lesser Black-backed Gull that flew among a flock of parakeets unsuccessfully trying to snatch a couple. That young gull clearly needs to hone its parakeet-catching skills pretty quickly, then pass that knowledge on to its mates. Peregrines have been known to take them and I once watched a Sparrowhawk go after one only to turn tail pursued by an angry emerald-green squadron. And that's the thing, have you checked out their bills? They are absolutely formidable.

As hole nesters, a lot has been said about them forcibly taking over the nest holes of other woodland hole-nesting species, leaving the original occupants homeless. I am so anti-parakeet that I believe this without question even though I have not seen it with my own eyes. Such is my prejudice. This is despite knowing of an old tree at my current patch, Wormwood Scrubs, where a pair of Rose-rings share the same tree with two pairs of Starlings and a pair of Great Spotted Woodpeckers – a veritable tenement block.

I confess that my dislike for them is probably very unreasonable and the logical part of my brain always questions me by asking: 'Yes David, but would you have felt the same way if you lived in Britain a thousand years ago during the time when Pheasants were being introduced? You happily consider them British now, don't you?'

It's a constant battle that I have with myself and there is never a winner. Will today's parakeet be tomorrow's pigeon? My guess is yes.

Airport Birding

'Late' has become my middle name recently. It has crept into my game at an inexplicable rate. I recently made it to Heathrow Airport to catch a flight to Belgrade to attend a conference on Long-eared Owls by the skin of Lady Luck's teeth. Had the flight not been delayed by an hour I would have missed it. Not good. After all that sprinting from the Heathrow Express terminal to the check-in desk, at least it gave me time to chill in the coffee shop and look out of the window with a nice cup of peppermint tea for a spot of airport birding.

I have always kept a mental list of the species that I have seen while taking off, landing and, where possible, strolling across the tarmac to and from the parked aircraft. I particularly enjoy guessing what species I will see first on touchdown, and there tends to be a steady cast of usual suspects. I'm invariably greeted by a corvid of some description, perhaps a lark or two, Starlings or perhaps some hirundines depending on the season and location of the airport. When you consider it, airports can be very interesting places to observe birds.

They are often situated just outside major conurbations, usually surrounded by marshland or bordered by the sea. This is great for attracting birds but can pose problems for aircraft in the form of potential bird strikes. Of course, not all airports are good news for the environment and wildlife. Building them involves the destruction of the natural habitat that they replace and all the associated repercussions on the natural life that was originally found there. However, some airport authorities are aware of this impact and do their best to encourage wildlife to their newly created environs.

I love scanning for birds from planes. Once airborne, I imagine that I am a bird looking down on the mosaic landscape unfolding beneath me. What warblers are in the hedgerows below, and are there any gulls or egrets on those bodies of water? Realistically, I don't stand a chance of seeing anything from 10,000 feet up, even if it was the size of an elephant. There is more chance of at least seeing gulls when you are flying at low altitude in a smaller aircraft. Recently, I remember being on a flight over the sea around the Channel Islands heading for Alderney and clearly identifying squadrons of Gannets going about their business. I even thought that I picked up a shearwater too.

Crucially, the area surrounding major airports is usually out of bounds, and since 9/11 security is often high with trespassing being strictly forbidden. Unofficial nature reserves are therefore inadvertently created. Heathrow is a classic example as its environs include reservoirs, balancing pools, woodland and grassland. As a juvenile birder I used to regularly breach the fencing around the now-defunct Perry Oaks Sewage Farm that used to lie right under the flight paths of roaring 747s. It was a legendary place that had a roll-call of rare birds which could compete with almost any site in the UK that you could name. Solitary Sandpiper, Lesser Yellowlegs and Aquatic Warbler were some of the delights to have graced this smelly place. In those days my illegal birding expeditions there would often be cut short by the sound of the sirens emanating from the local constabulary's panda car and the long arm of the law turfing me out.

Grasslands are an obligatory component of an airport and it's a well-known fact among birders that these areas are always worth a good scan, especially if the airport is on an island or headland. Lapwings, Golden Plovers and recently a Dotterel have all been found wandering the airfields at Heathrow, but the airstrip on St Mary's Airfield, Scilly Isles, is legendary. I've personally seen a Short-toed Lark, Richard's Pipit and Lapland Buntings but that's only the tip of the iceberg as there have been many great birds discovered there, including such beauties as American Golden Plover, Buff-breasted Sandpiper and the island's first Sociable Plover. It's a bit of a weird concept looking for rarities at an airport as it's almost as if they are travel terminals for both humans and birds.

As I stared out at a beautiful blue sky at Heathrow watching for raptors and any avian movement on terra firma, I recalled previous, more successful times at other airports, such as at Inverness Airport, where I was lucky enough to watch a skein of Pink-footed Geese fly over. Landing at an airport near Kuhmo in eastern Finland, the first birds I saw were not the usual corvids but Curlews and a solitary Jack Snipe that towered out of its grassy hiding spot at the approach of my plane. The further afield you go the more varied your airport birding will become. When I landed in Cancun, Mexico, a few years ago, I witnessed the incredible sight of many thousands of grackles flying in to roost in a thick jungle that adjoined the airport. I later found out that they were Greater Antillean Grackles – a nice tick for me.

My time was up as my flight was being called. I wondered what my first bird would be when I arrived in Belgrade. How many times have you stopped to observe the birds at the airports that you have travelled through? If you haven't tried it before, then spare a moment to look around. You just might surprise yourself.

The British Isles and Ireland

Aberlady Bay, East Lothian, Scotland

Bonnie Scotland has barely figured in my column over the years, to my eternal shame. I was acutely aware of that fact one recent autumn as I sat on an Edinburgh-bound intercity. However, it was not the Scottish capital that I was hurtling towards but a certain beauty spot called Aberlady Bay in East Lothian. Officially, Aberlady is a town, so my urban authenticity was intact, but it was the thought of hanging out by the coast away from the madding crowd that really attracted me. Both seeing and hearing waves crash against an ancient shore sending Sanderlings skittering was not something that I saw with any regularity from within my urban landlocked west London local patch. My loose plan was to explore the East Lothian coast, of which Aberlady Bay forms a section, to see what I could find. One of the key attractions in this part of the world are the enormous flights of up to 30,000 Pink-footed Geese that pass over the bay to and from their feeding and roosting sites. Aside from the Pink-foots, there had also been the autumnal arrival of some 2,000 to 3,000 Barnacle Geese. That was a sight that I did not want to miss, especially knowing that their numbers were peaking at the time of my visit.

In 1952 Aberlady Bay, some forty-five minutes drive east from Edinburgh, had the honour of being designated as Britain's first local nature reserve. It covers an area of 1,439 acres, two-thirds of which is actually under water, below the tidemark. So as you can imagine, there are plenty of mudflats and tidal sand providing a magnet for wading and waterbirds. Indeed, it is the winter period that supports the best birding, with the chance of a Short-eared Owl, Merlin or Peregrine hunting over the bay that is filled with Knot, Ringed Plovers, Bar-tailed Godwits and other waders. I was staying in Aberlady and the short walk from my hotel across the road to the bay transported me into a world filled with the hypnotic wild cries of Curlew and Oystercatchers. Hearing those evocative calls never fails to excite me. The tide was not quite in, but a scan across the vista rewarded me with nice views of Wigeon and Red-breasted Mergansers; stalking the mud were Redshanks, Grey Herons and several Little Egrets. Nowadays, we struggle to remember just how rare those egrets once were in the UK.

I was guided for part of my stay by the Scottish Ornithological Society's Dave Allen, a garrulous man who gladly offered to take me around his stamping grounds in the company of accomplished watercolour artist and birder Darren Woodhead. Over the next couple of days we explored a great deal of the coastline, starting with the bay on my hotel doorstep, my surrogate local patch. Variously swimming in the coastal waters and feeding on the shoreline were a fair number of Shelduck. Up to 400 spend the winter each year, while good numbers of Fieldfares move in to feed on the Buckthorn berries from January. The guys explained that summertime here is pretty good, too, as you have a fair chance of discovering breeding Grasshopper Warblers as well as Lesser Whitethroats – the latter a scarce bird this far north.

I really enjoyed birding at Gosford Bay further west of Aberlady. It seemed to be a grebe Mecca because during August the area attracts the highest counts of Red-necked Grebes in Britain with up to fifty birds present. Supporting its grebe credentials are the winter flocks of upwards of eighty Slavonian Grebes that grace the waves. During our visit we managed to pinpoint a few Slavs mingling with some early Long-tailed Ducks. It really is a super place for seabirds and seawatching during the winter, with the three divers, auks, and Velvet and Common Scoters all on show. Terns also gather in the autumn, including the occasional rare Roseate Tern.

Our focus switched to passerines when we pulled into the western car park at Longniddry Bents. The coastal scrub screamed 'rarity haven' to me and I immediately got to work scanning the terrain for any signs of wayward waifs. Dave got me to calm down and instead look for Fieldfares and other winter thrushes, Blackcaps and Bullfinches. It certainly did look like a very interesting site but all we managed were a few Goldfinches. The coastline east of Musselburgh was particularly rewarding and afforded stunning panoramas of the entire bay. We had amazing views of a group of Long-tailed Duck bobbing on the sea close to the coast. Alongside them were Eider and more Red-breasted Mergansers, while an army of Black-headed and Common Gulls swooped and swirled over their heads. Distantly, the guys picked out a lone Bonxie patrolling the horizon and an equally distant small group of Little Gulls.

I spent a very relaxing couple of days in great company enjoying some leisurely birding; that for me is the best birding. Why rush around when you can stroll or even sit? Let the birds come to you is what I say. My favourite moment was when we joined a group of Scottish Ornithologists' Club members waiting near the SOC HQ in the pre-dusk light for the evening flight of Pink-feet. They certainly did not let us down. Thousands headed over us in multiple skeins, yelping like a vast pack of small dogs. It was exhilarating. So, I didn't quite make it to the Scottish capital. That is to come. But I must say, the East Lothian coast will take some beating.

Alderney, Channel Islands

Picture the scene: the sun is blazing down, time has practically stood still and I'm stretched out without a care in the world on a seaside grassy knoll on the quiet and extremely picturesque island of Alderney. I'm watching my second Honey-buzzard of the week drift over the sea in azure skies from nearby and clearly visible France, a mere seven miles away. You must be wondering what I was doing lounging on Alderney instead of eking out birdlife in some inner-city oasis. Well, I was on Alderney at the behest of the Alderney Wildlife Trust and the island's tourism office, both of which were keen for me to experience the wildlife to be found on this, the most northerly of the Channel Islands.

So why was I writing in my urban birding column about a Channel Island that most birders haven't been to and that the majority of the general public think is part of the Orkneys? In truth, it was never my intention to write about the island. My original plan was to be part of the wildlife week organised by the Alderney Wildlife Trust, which involved me leading a couple of walks (including an urban walk through the streets of St Anne, the main town on the island) and giving a talk on urban birding. Now, Alderney is not a big place. At just over three miles long by a mile and a bit at its widest point, you could easily walk around it in a few hours. But my interest was raised because the island feels like what suburban Britain must have felt like in the Fifties; quiet, neighbourly (people still leave their doors open) and residential.

I quickly learnt that Alderney is an island of contrasts. Despite its obvious beauty remnants of the wartime Nazi occupation could be seen, with the occasional fortress and bunkers peppering the coastline. Most had either been refurbished or reclaimed by nature, with breeding Meadow Pipits, Dunnocks and Wrens claiming them as their own along with the Swallows that bred in some of the bunkers. I was staying in a gorgeous old building on the north-east coast at Longis Bay with two elderly cats, including a deaf, moth-eaten moggie that had reputedly not left the building for several years. The latter cat had a habit of spending the day sprawled over the arm of the living-room sofa. One afternoon I was lying on the sofa taking a nap when I was rudely awoken by the flippin' cat jumping on my face, digging its paw into my eye as it made the final leap to the arm of the chair. I shouted and cussed at the cat, but of course it fell upon deaf ears.

Alderney also only has two birders in a population of around 1,200. Mark Atkinson, the local bird recorder, and Alastair Riley, a birder and very talented artist, were my guides and confidants. Over the ensuing week I adopted a local patch around the bay and the adjacent gorse-strewn Longis Nature Reserve. I discovered the island's only breeding pair of Little Grebes on Longis Pond and watched ubiquitous Meadow Pipits launching themselves into the air in song flight, while the prolific Blackbirds flew low across the common like all-dark Ouzels. I also found a singing Dartford Warbler in the gorse at 'the Odeon', a prominent German-built tower on the brow of the hill overlooking the reserve.

The Dartfords were down to perhaps five pairs after having suffered horribly at the hands of a recent harsh winter. The island's population of Stonechats had been extirpated by the freeze, and the guys could find no sign of the tiny Zitting Cisticola population, though there was a single breeding pair of Serins residing close to where I was staying. It is Alderney's close proximity to France that made the place so exciting for me. Honey-buzzards and Montagu's and Marsh Harriers nested in good numbers literally across the water in France and were prone to daytripping to Alderney on a regular basis. Hence the four Honey-buzzards I saw during my stay, including one that I watched flying over the sea, circling around my patch of the island for 20 minutes before drifting back to France.

Exploring the island, I was surprised to find that Blue Tits were an extreme scarcity and that Jackdaws numbered only a few pairs to be found in the south. Buzzards were commonly seen, and closer to the ground the nationally rare Glanville Fritillary was a frequent sight. Working the island is a must especially around the airport area in the south-west. The cliffs at Les Etacs and Ortac Rock were covered with nesting Gannets – a truly impressive vision. To complete the nautical vibe, I did a boat trip around Burhou island, which lies west of Alderney. I saw many of the area's 200 pairs of Puffins (some of which are viewable from the Wildlife Trust's Puffin Cam) and learnt that the islet was home to a sizeable colony of Storm-petrels.

Of course, the possibility for vagrants must not be ignored, and the island's bird list is impressive by anyone's standards with juicy goodies like Nutcracker and Trumpeter Finch. But with only two resident birders, they've got a serious job on their hands and need all the help they can get.

Suburban island life was great, but for all you city lovers out there, normal service will be resumed shortly.

Belfast, Northern Ireland

Take me to a city and I'll show you a bird. That has been one of my mantras since before I actually understood what the phrase meant. I must be the only birder on the planet who actively relishes the prospect of birding in concrete jungles. Having cut my teeth in London I have noticed that most cities around the world seem to be designed using the same blueprint. They all have a combination of parks, botanic gardens, cemeteries, lakes, rivers, marshes, sometimes shorelines and plenty of sky. All these things together mean one thing: birds.

I decided to put my mantra to the test to see if I could find birds in the cities that you may not think are birding venues. I thought I would explore places where initially the only visible evidence of avifauna of any kind might be found fried, slapped between a bread bun and sold in dodgy fast-food joints.

With bins, scope and iPod in tow, my journey into the world on our doorstep started with Belfast. While planning this visit, my questions about inner-city birding sites were met with incredulity from several of the Northern Irish birders that I contacted. Why go birding in Belfast, they cried, when you can go to Lough Neagh, Rathlin Island or Copeland Bird Observatory and see some proper birds?

Well, when I stepped off the plane at George Best Belfast City Airport for my two-day visit I was heavily laden with a list of sites, a freshly acquired cold and an insurmountable excitement. After checking in with my guide and B&B owner Margaret Adamson, I settled down for a good night's kip. The next morning found us at our first destination – Ormeau Park – apparently the biggest and most species-rich park in the city. Strolling around this very pleasant wooded site resulted in plenty of Treecreepers and the common tit species, while overhead roving parties of Mistle Thrushes 'chacked' with smaller numbers of Song Thrushes and Redwings in their midst. I had never seen so many migrant Mistle Thrushes before – they were all over the place throughout my stay.

We journeyed east across the city to the once-notorious Falls Road to visit the beautiful Belfast City Cemetery and to satisfy my curious fascination with burial places. I just love cemeteries. This one was huge: more than 100 acres consisting of mixed woodland and scrub interspersed by some surprisingly ornate tombs and headstones. Although I did not see much other than hoards of roving Mistle Thrushes, I could see that the place had a lot of promise. In fact, it is said to be the most bird-rich area, in terms of breeding species, in the whole of the city. It was the sort of place that I would definitely adopt as a local patch if I lived in Belfast. Literally adjoining the cemetery to the west was Falls Park, where there is the possibility of seeing Grey Wagtails and Dippers frequenting the stream that runs through it. We never got that far because time necessitated a trip west across town to Belfast's jewel in its crown, the RSPB's Harbour Reserve. This site on the shores of Belfast Lough is a surprisingly small, square-shaped lagoon sandwiched in the middle of a dockland industrial estate and beneath the flight path of the nearby city airport. Despite this, it has garnered an amazing tally of rarities. Nearctic waders like Pectoral and Semipalmated Sandpipers are practically annual and as for gulls and terns, it was more a case of what hasn't turned up. I was particularly impressed with the tern rafts that during the summer are populated by both Common and Arctic Terns. And the past two years have seen a couple of pairs of ultra-rare Roseate Terns prospecting.

The following day I discovered Kinnegar just down the road and down the shore from the RSPB reserve. Here I counted Red-breasted Mergansers, Great Crested Grebes, a few lingering Sandwich Terns and a sole pale-bellied Brent Goose. For the rest of the day I checked out a few other sites, ending up at Belvoir Forest Park on the southern edge of the city. Belvoir (pronounced 'Beever') is a glorious tract of mixed woodland that seemed to be stuffed full of Jays. Apparently, it holds breeding Goshawk, but my day was crowned by seeing a Red Squirrel taking nuts from a feeding station inches from my face through a window of the RSPB's Northern Ireland Head Office.

I'm not a stranger to Belfast as I had birded in the city some fifteen years previously. Back then the atmosphere was decidedly edgy. I remember visiting a mate who lived in a Catholic area, and he went ballistic when he opened the door to see me in a green army birding jacket. He vociferously ordered me to take it off, as he was afraid that I would be mistaken for army personnel.

Belfast today is a very different place. It's cool, it's home to the delightful Christine Bleakley from BBC1's *The One Show*, and its people are unbelievably friendly and courteous. But perhaps most importantly, it is a surprisingly good urban birding venue.

Bradford, Yorkshire

Bradford, in the foothills of the Pennines, is famous for its once-thriving textile industry and, latterly, its decent curry houses. Things certainly got a bit hot and spicy on my arrival night when I was rudely awakened out of my deep slumber by an unnecessarily loud hotel fire alarm going off. A drowsy call to reception confirmed that there actually was a fire (someone had set their pillow alight – don't ask me how), so at 1.30 a.m. I was standing outside on the cold street with all the other sleepy hotel guests in various states of undress. A few were a bit disgruntled and were using some choice spicy language while we waited for the fire brigade to do its thing.

My hotel was situated on Hall Ings in the centre of town, and whilst aimlessly gazing around the darkened environs I noticed a few rowans and cotoneaster trees liberally scattered along the street. I later learnt that these trees were magnets for Waxwings. Once the local birders got wind of the news that winter flocks of these gorgeous birds had made landfall in cities like Aberdeen and Edinburgh, a Bradford invasion was generally on the cards.

After my disturbed sleep I managed to drag myself out of bed for a pre-dawn visit to Bowling Park – ten minutes drive south-east of my hotel. At around fifty acres, this municipal park had some interesting wooded and scrubby areas that immediately triggered my inquisitive nature. My instinctive hunch was proven right when I noticed mixed flocks of Redwings, Blackbirds and several Mistle Thrushes in the trees busily stocking up on berries silhouetted against the moody early-morning sky. Chaffinches, various tit species and a couple of Goldcrests called as I speculated on the variety of warblers that must breed in the area during the summer.

There was a large cemetery adjoining the park to the east, and being a lover of such places I made a beeline towards it via some playing fields that were covered with Common and Black-headed Gulls. However, time got the better of me and I had to abort the mission and head to nearby Bingley to meet Shaun Radcliffe, Chairman of the Bradford Ornithological Group and my guide for the day.

The city of Bradford itself is quite industrial and generally lacking in many areas of green space. The metropolitan area, however, is a very different story. It encompasses woodland, fast-flowing rivers, boggy areas, standing water, miles of stone walls, reed beds and, of course, seemingly endless expanses of moorland. Shaun and I started our look at Bradford by literally viewing the city from the uplands of Rombalds Moor. This moor is part of the famous Ilkley Moor that among other things is eternally connected to the Brontë sisters. Shaun gave me the list of species to be found on the moors, which read like a compendium of the birds you would expect in this kind of habitat. But that close to civilisation? I mean, how many cities are there in Britain where you can observe a Red Grouse surveying the scene from the top of a boulder with a glinting cityscape as part of its backdrop?

The surrounding areas of Bradford are extremely picturesque particularly around Bingley – a town where Waxwings are also sometimes expected during the winter and where the River Aire is replete with Dippers and Otters – and around Keighley, pronounced Keith-ley just to confuse visiting southerners like me. In Bradford itself we visited a secret site to watch a Peregrine perched on the side of an old mill funnel, then Shaun replaced the sack over my head, spun me around a few times and took me to an ultra-secret urban hotspot. We arrived at a nameless street and randomly knocked on a resident's door. After a bit of negotiation that I was not privy to, the occupants took us around the side of the house to their back garden, which was basically a short veranda that backed onto a reasonably sized deep lake.

Red Beck Mill Pond is completely encircled by back gardens and therefore totally invisible from the street. In winter it regularly attracts diving ducks like Tufted Duck and Goosander, and the homeowner who we were talking to spoke of resident Kingfishers. This was a real urban oasis. Nearby in an industrial estate at Wakefield Road, around 100 Lapwings were roosting on the roof of a factory. Shaun said he had counted a similar number of rooftop-loving Golden Plovers using the estate a few days previously. Our street-by-street search drew a blank but did receive attention from a police officer scoping us from an unmarked car. He was a little suspicious of Russell, my photographer, who was stalking around looking for the plovers with his long lens. Apparently, there was a very security-sensitive police building on site.

Perhaps the most well-known urban spot is Raw Nook Nature Reserve to the south of the city and beloved of Bradford's migration watchers. The site seems to pull in the avian travellers traversing the Pennines, scoring good counts of the more common migrants passing over, plus a few scarcer ones like a male Snow Bunting. In the summer the site also boasts the area's only breeding Lesser Whitethroats.

Bradford was far more diverse as an urban birding venue than I ever imagined and is a city that I would return to at the drop of a flat cap.

Brighton, Sussex

My ongoing urban birding odyssey to the cities of Britain and Europe has been a massive learning curve. I have encountered many interesting species, some of which I had never expected to see, and I have befriended some very knowledgeable and passionate urban birders. But what never fails to amaze me is the shocking number of cites, some very well known, that are seemingly completely underwatched.

One such place is Brighton (or Brighton and Hove to give it its full official title). Recently given city status, this south-coast conurbation lies between some pretty interesting birding spots with places like Pagham Harbour not a million miles away to the west, and Seaford, Cuckmere Haven and, of course, Beachy Head to the east. I can understand why birders overlook Brighton.

My history with Brighton is less to do with birds and more with the romanticism that the place evokes. Yes, I once had a love interest there (although, alas, she wasn't that interested in me), and I saw and worshipped *Quadrophenia* and toyed with the idea of spending a summer of my youth bumming around the seafront. Instead it ended up just being a succession of day trips. The first time I ever raised a pair of binoculars in a bird's direction while in Brighton was during the early Nineties when I was in town for a conference. I remember walking out of my seafront hotel on a squally grey morning on the cusp of spring for a pre-breakfast seawatch. Almost immediately I saw an adult Mediterranean Gull sail past over a fairly angry-looking sea in the loose company of some ubiquitous Black-headed Gulls. Later, I took a drive to Seaford where I discovered some inshore Guillemots, a Black Redstart and several early Wheatears.

This time around, on a cold winter's morning, I took on the role of urban explorer with no real plan but to walk around and see what I could find. I am very much a believer that anything can turn up anywhere at anytime, so I initially stationed myself along the promenade and stared out to sea, scanning for seaduck and gulls. I was approached several times by members of the public, who asked if I was a twitcher and had come to see something rare. My mini-seawatch resulted in a couple of Gannets, a raft of twenty-two drifting Great Crested Grebes and numerous seagulls of the usual common species. I think it's well worth giving the sea a scan almost anywhere along the seafront, while nibbling on your candyfloss, because many birds seem to commute over the sea past Brighton en route to no doubt quieter environs.

Brighton Marina was another interesting, if incongruous, place to explore. It was racked full with a panoply of neatly moored boats ranging from small, simple fishing craft to luxury yachts that looked more like floating Bentleys, judging from their car-like dashboards. Encircling the marina was a fortress-like sea wall that you can walk along, and it perhaps offered the best seawatching opportunities in town. I watched a couple of Rock Pipits entering the crab pots next to the local fishing boats along with a few House Sparrows and a host of Starlings, the archetypal Brighton bird – more about Starlings later. The marina is also well known as a good place for wintering Purple Sandpipers and Black Redstarts. A quick examination of the massive lumps of concrete liberally dumped in the eastern corner resulted in a pair of each of these engaging species. I was particularly pleased to see the Purple Sandpipers as I hadn't seen one in years.

I say that I didn't have a plan, but in reality my whole day was built around seeing the Starlings coming in to roost on Brighton Palace Pier. As I had a good hour before the spectacle began, I journeyed further west along the coast to Saltdean on the edge of Brighton. There was a nice seafront walk that ended within sight of some impressive cliffs, which played host to wheeling Fulmars. On the rocky coast I scoped an array of gulls, including a fine-looking winter-plumaged Mediterranean Gull and three Little Egrets.

It was late afternoon, so soon I was taking my place along the front by the pier among assembled onlookers for the Starling roost grand finale. Within minutes we were all watching at least a thousand of these great birds flocking over the sea. Satellite groups would fly in above the main mass only to be sucked into the mother flock as if they were caught in an irresistible traction beam swallowing them in like water down a drain.

Now I've seen Starlings coming in to roost a thousand times before, but I can tell you that I would watch them a thousand times again – it is just so mesmerising, calming, humbling and peaceful. I looked around me and saw ordinary people watching with their mouths open in wonderment. They smiled and gushed as the flock swarmed and shimmered, cutting animate shapes. It was as though they were all watching a firework display.

I made my way onto the pier itself as the flock, which by now must have numbered at least 30,000, whooshed close by before swinging back out to sea. A battery of camera phones flashed each time the birds approached. Where else can you get this close to such a natural wonder in an urban environment?

Night fell on my day trip to the seaside and I returned home with the sound of chattering Starlings still ringing in my ears.

Bristol, Avon

Bristol is a magnet for people professionally involved in natural history whether they are in front of the camera or behind it. Many gravitate especially to the BBC's Natural History Unit, home to the crème de la crème of wildlife film-making. The NHU's existence in the city has led to Bristol's other moniker – Green Hollywood. As a consequence of this unusually high percentage of naturalists, the city and its wildlife are perhaps the most studied in the UK and maybe even in the world. Did you know that until an outbreak of mange in 1996 Bristol had the biggest population of urban Red Foxes in the world, and that there are currently more Badgers there than in any city in the UK?

Shining accolades indeed, but what of the urban birding? It is a fairly well-known fact that Bristol is a brilliant place to observe urban Peregrines, since they returned to breed in 1990. Personally, I know the city fairly well due to my regular visits to the NHU, but until recently my knowledge of the birding scene was embarrassingly paltry. It's time to introduce local naturalist and photographer Sam Hobson. Once a regular observer on Tower 42 in central London, he has since moved to Bristol and is now studying the natural life on Brandon Hill, apparently one of Britain's oldest parks, situated right next to the city centre. His invitation to show me around his patch and to sample the urban delights of his adopted city was one thing, but to be involved in the launch of the Cabot Tower Bird Study Group totally clinched the deal for me.

Cabot Tower stands at the pinnacle of Brandon Hill and was erected in memory of John Cabot, the Italian navigator and explorer widely believed to be the first European to clap eyes on the New World in 1497. So, on a cloudless pre-dawn September morning during a belated brief Indian summer, a group of us led by Sam climbed the spiral steps within the tower to greet the glorious dawn and great panoramic views of the slumbering city. At 105 feet, the tower was by far the

highest point around, and we had unrestricted views of the
city and the picturesque Avon Gorge. As the blood-red sun
started to break the horizon the waves of Meadow Pipits,
Chaffinches and Siskins began to sweep past us, sometimes
seemingly within touching distance. For the next couple of
hours parties of 'alba' wagtails periodically passed through,
while beneath us in the canopy of the park Chiffchaffs,
Greenfinches and a solitary Goldcrest caught our attention.
We saw a couple of Peregrines distantly perched on a church
tower and noticed quite large numbers of Jays. We also
watched small parties of urban Ravens as they tumbled and
playfully headed across the morning skies. This big corvid is
a species that I am still not used to seeing against an urban
backdrop in Britain. The previous week, Sam had watched a
single flock of twenty-one birds drift over the city. What an
incredible sight that must have been.

As I mentioned earlier, Cabot Tower stands within
Brandon Hill, a pretty wooded park that also contains a
small Avon Wildlife Trust reserve. This green oasis is clearly a
bit of a migrant trap, with scarce goodies like Wood Warbler,
Yellow-browed Warbler and, more recently, a Common
Redstart firmly on the site's list. Sam even found a Kingfisher
here, far from the nearest body of water. The more usual
avian inhabitants include resident Treecreepers, Nuthatches
and during the summer an array of common warblers
including Garden Warblers. I experienced the dubious
pleasure of having a rather bold Grey Squirrel run up my
leg while I was otherwise engaged looking up at a Great
Spotted Woodpecker. Sam did not convince me that it was
only looking for nuts.

I was naturally drawn to his patch as I saw great potential for some good urban birding. However, Sam did manage to draw me away to explore some other spots. For
example, Bristol Downs, an open park on the northern
side of the city, seemed like an interesting place. Back in
the Thirties the area boasted breeding Hawfinches and

Cirl Buntings, and nowadays it is a good venue for Lesser Whitethroats, Starlings, Bullfinches and, during the winter months, Redwings. We clocked a Common Buzzard being mobbed by a Sparrowhawk and little else as we sweated under an insanely hot autumnal sun. On the edge of the site was the Avon Gorge replete with its unique flora, such as the Bristol Onion and Bristol Whitebeam, as well as Peregrines.

After lunch we were joined by the delightful broadcaster Ed Drewitt, who had organised for us to join a cruise down the Avon through the centre of town. As it was a gorgeous day, the boat was filled with day-trippers and for us it was standing room only. Ed eagerly pointed out the various areas that could hold the attention of visiting birders. Castle Park right in the heart of the city was the key spot as Peregrines are always a possibility. As if on cue, I looked up to witness a male and female Peregrine high overhead attacking a feral pigeon in unison. It was such an amazing show that even some of our fellow passengers were marvelling at this most primeval of spectacles.

I had an incredible time discovering the flip side of Bristolian life, and doing it in baking sunshine certainly helped.

Cambridge, Cambridgeshire

As we all know, Cambridge is a university town steeped in history, the streets of which are riddled with millions of bookish students riding bikes. When I came to visit recently, I could not stop myself from being captivated by the architecture of the impressive university buildings or from feeling conscious of my profound lack of education. But here's a sobering thought: Cambridgeshire is the least wooded county in the least wooded country in Europe. That thought certainly had me scratching my chin as I stood with Peter Herkenrath, the chairman of the Cambridgeshire Bird Club, by an icy stretch of the River Cam. He dropped that fact on me while we were in the centre of Cambridge counting some chilly looking Black-headed Gulls that were variously congregated on the moored barges or on the frozen river itself. Cambridgeshire's lack of woodland appears to be verified by the Woodland Trust. Thankfully, there are still tracts of woodland in and around the city rich enough in their biodiversity to sustain healthy populations of wildlife.

I was still pondering tree cover as Peter and I walked through Jesus Green, just north of the city centre and close to Jesus College. Despite the multitudes of people who play football, rounders and other sports there, especially during the summer months, the urban birding is still good. The stretch of the Cam there supports the exotic Mandarin Duck and Common Terns that gracefully hawk over the water's surface during the summer. A few pairs may breed on a nearby flat roof but this has never been proven. Walking around the Backs, the imaginatively named characterful gardens and woodland at the backs of the colleges, was a pleasant experience. Much of the area is private although there is some limited access. Aside from breeding Coal Tits and Goldcrests, it is the only place in the city where you can find the Nuthatch – a rare species in Cambridgeshire, so rare that we could not find any. The college buildings themselves harbour the occasional Black Redstart, while Grey Wagtails breed along the ever-present River Cam.

A couple of interesting species have shown up in the city over the years, such as the Leach's Petrel that incongruously flew past Castle Hill to the north of the city centre. Perhaps Cambridge's biggest ornithological head-turner was the pair of Moustached Warblers that nested in 1946 at Cambridge

Sewage Farm. Darker and crowned with a stronger supercilium and a more rufous tone than their Sedge Warbler cousins, Moustached Warblers are partially migratory marshland denizens of parts of eastern Europe and southern Asia. After the surprise discovery of the Cambridge birds they were watched for hours on a daily basis for a few weeks by a group of birders that included a roll-call of the most eminent ornithologists of the day. Many of the observers made quite detailed notes that all seemed to point towards this most unusual vagrant. The record stood for years despite the slight murmur of disagreement from a few ornithologists who felt that the birds were no more than aberrant Sedge Warblers.

After listening to the story and reading the original accounts I too was convinced that Britain had indeed received a visitation from this unlikely species. I was therefore shocked to hear that the record was finally thrown out by the British Birds Rarity Committee (BBRC) in 2005. It was for good reason, as it appeared that it was a case of mass hallucination. Closer scrutiny of the notes and descriptions had found a number of inconsistencies that helped to rule out Moustached Warbler. The BBRC sleuths found that the case for Sedge Warbler was significantly strengthened when it was realised that the observers had based their prognosis on the field guide of the day's illustration of Moustached Warbler that only showed its underparts – hardly useful for identification. Crucially, the 'one last thing' moment, to coin the great Columbo, was that observers universally remarked that the birds had pale legs. This is a Sedge Warbler feature – Moustached Warblers' legs are dark. The case was solved. The moral, I guess, is that when on a twitch we must always ask ourselves the question: why is the bird that we are watching what they say it is?

Sadly, the sewage farm does not exist anymore, so we couldn't go with heads respectfully bowed to visit the scene where this incident occurred. Peter was a bit sad that this record ultimately did not stand as it would have been a truly remarkable occurrence.

We ended our day at the delightful Milton Country Park, very close to where the infamous sewage farm used to be, and home to winter parties of Gadwall, Shoveler, Teal and other common waterfowl species. During the summer the expected common breeding warblers are readily seen, but it is also a good place to find a few regularly nesting Cetti's and Grasshopper Warblers plus a couple of pairs of Nightingales. We were watching a Water Rail feeding at close range around an unfrozen water margin when a hitherto cryptically camouflaged Snipe suddenly jumped up and momentarily slanted its body sideways towards the bewildered Water Rail flashing the rusty feathers of its splayed tail. They crossed paths with the rail resuming its foraging and the Snipe sitting down to melt again into the wintry vegetation. The whole episode seemed to last a nanosecond. I had never seen that kind of behaviour before, and I will never forget it.

Croydon, Surrey

I've been to some pretty weird places in my search for urban birds, such as picking my way across derelict land looking for Northern Wheatears and observing migrants in between sipping tea sitting on the top floors of inner-city balconies. I've even birded while lying in bed overlooking suburban back gardens, but so far nothing has been as challenging as the thought of going birding in Croydon.

Now before you start thinking that I have finally hit a birding brick wall, let me quickly state that the reason for my trepidation was not for the potential lack of birds. It was because of the sheer negative reputation this place has. I had to wrestle with my own ingrained prejudices about Croydon. Was it nothing more than a load of skyscrapers jutting out of bleak, rain-stained concrete streets, inhabited by chavs who relish hanging around in equally fake-looking shopping centres? Was Croydon in south London or was it a province of Surrey with its own queen? I was confused. Even checking the Internet resulted in finding a website written by a Croydon-based birder deriding the birding opportunities. How much more discouragement did I need?

But I now know better. Croydon is like any other large town; it has nice bits and some not-so-nice bits. The people I encountered were like any others I have met on my travels – just ordinary people. It's an ancient place that was in Surrey until the skyscrapers came in the mid-Sixties, when it became a London borough. The Croydon name derives from the Saxons and is thought to mean either 'Crooked Valley' or 'Saffron Valley' – which to my mind is what Croydon should have been christened. Take That recently shot part of their latest video there and it's also the place where Kate Moss first stepped out of her knitted booties and into stilettos. On a personal level, one of my closest friends lives in the city, and there are plenty of birds here too.

To prove that last point, when I arrived in wonderful Saffron Valley I strolled literally minutes away from East Croydon Station into the nearest green patch, Park Hill, a small municipal open space. After parking myself on a bench, I was soon enjoying prolonged views of Goldcrest, Great Spotted Woodpecker and a gaudy, vociferous Nuthatch. Unbeknown to me, a Firecrest or two had been lurking there for the previous few days. I was surprised and taken in by this small park's avian richness, all within sight of the city skyline.

Giddy from my brush with nature, I brought myself down to earth with a thud as I walked through the town centre.

Initially, aside from a few high-flying gulls there was nothing on offer other than the obligatory feral pigeons. But I have been told that if you are diligent you may see the resident pair of Peregrines that sometimes flash across the skyline. Last summer there were potentially two pairs of Grey Wagtails and a Black Redstart on territory and, in addition, there is a reasonable Pied Wagtail roost during the winter.

Bright and early the following morning, I found myself wandering around the National Trust-owned Selsdon Wood Nature Reserve. I was lured to this area of suburban woodland and open meadows by the prospect of catching up with Marsh Tits, as this reserve is one of the main sites for this ultra-scarce species in Surrey. I was unlucky in my quest, but I did manage to discover, although not see, a calling and drumming Lesser Spotted Woodpecker – a species that I barely see these days. Many of the familiar woodland birds were quickly added to my tally, including several Great Spotted Woodpeckers and calling Treecreepers. This site has been somewhat under-recorded recently and would definitely benefit from some regular coverage.

I was clearly enjoying being a suburban birder, so I made the short drive over to Riddlesdown, an area of downland that is home to several pairs of Skylark, Meadow Pipit and Yellowhammer, and is one of the few places in Croydon where all three nest. As I stood looking over the gentle valley, I imagined the possibility of seeing raptors coasting by, so a Peregrine that I noticed soaring over my head in the beautiful blue sky came as no surprise.

The final part of my whistle-stop introduction to Croydon birding was a trip back into the heart of things with a visit to South Norwood Country Park. Despite being in the middle of an urban area and thus heavily visited, its mix of scrub and wet meadow looked very appealing. With over 170 species recorded to its credit, including Long-eared Owl, Great Grey Shrike and Twite, this site is without doubt the hotspot of Croydon. As I watched a male Reed Bunting staking out its territory, a passing local birder told me about the persistence of a Jack Snipe in the wet fields near the lake. The snipe eluded me, so instead I had to make do with watching some Shovelers chilling in the company of Canada Geese. I left feeling that the country park would make for a very rewarding patch.

No matter what people may say or what you may have previously thought about the Saffron Valley it is definitely worth several repeat visits. Whether you class it as London or a part of Surrey, with over 150 parks and open spaces in the area, most of which are at best underwatched, avian riches await those who seek them.

Derby, Derbyshire

Derbyshire Wildlife Trust's Nick Brown is a dead birdwatcher. Before you recoil in horror, I must stress that he himself is very much alive and kicking. You see, Nick delights in creeping around on the ground directly below Derby Cathedral looking for the mangled remains of victims discarded by the city's nesting Peregrines. The birds perch on the gargoyles, or to give them their correct name, 'the grotesques', to dismember their prey, resulting in the occasional discarded body part. Ghoulish as it may sound, I can see the excitement in this branch of ornithology. Interestingly, in New York there are dead birdwatchers who collect the nocturnal migrant victims of collisions with the Empire State Building.

Over the past five years, Nick has amassed a dead list of more than fifty species, ranging from the obvious pigeons (of Wood and feral varieties) to birds that you wouldn't imagine a Peregrine tackling, such as Waxwing and two Corncrakes – one of which was a bird that emanated from the Nene Washes reintroduction scheme. Nick's main gripe was that when he comes to check the ground for the latest kills, it has often already been swept clean of unsightly bodies by the vicar. Indeed, when the Peregrines first took up residence on the cathedral the presence of random bird heads scattered around was originally seen as being some sort of satanic ritual.

Derby Cathedral has an impressive tower which was built in *c.* 1535 and is the second tallest in Britain. It felt like the tallest tower in the world when I climbed the ancient, curling steep staircase that led to the bell ringers' room and eventually onto the roof. Nick and I were joined by Derbyshire Council's Nick Moyes, who works for Derby Museum and who facilitated our access. He was the man responsible for setting up Derby's internationally famous Peregrine webcam, whose nerve centre was in the bell ringers' room. We watched four young chicks in the nest on the live feed and viewed some amazing recorded footage, including a sequence where one of the adults dispatched a live Woodcock under the cover of darkness. I could see why the website had over 500,000 hits.

Afterwards, we headed to the roof and enjoyed great views across the whole of the city in glorious sunshine. By now the bell ringers had come in and were knocking out a tune with vigour to attract the congregation. Peering over the wall onto the top of a grotesque I was shocked to see the fresh remains of an adult Little Grebe. On another grotesque were the remains of a couple of Golden Plovers. Nick Brown once found the head and leg of a Scandinavian ringed Arctic Tern. There's even footage of a bird feeding its young with a Brown Rat. I never knew that Peregrines plucked prey from the ground.

I could have spent all day at the cathedral but there is more to Derby than its famous Peregrines. Due to time constraints we had to cut straight to the chase and walked down a portion of the Derwent just north of the cathedral. The river attracts Dippers during the winter, resident Grey Wagtails and nearby breeding Goosanders. There have even been Otters reported. The wooded parkland that we were walking through also looked interesting, but a fun run noisily taking place that morning put paid to us seeing anything other than the obligatory Robins, Blackbirds and tits.

Aside from visiting the cathedral, my main memories of Derby are the time that we spent in the Pride Park area; home to Derby County's football stadium. On the river that flowed nearby we watched a Sand Martin colony that had ingeniously created nesting sites in the holes of the metal girders which flanked the riverbank. Apparently, at least thirty pairs breed here. While watching a large Pike sedately swimming past, glistening in the water, a dazzling Kingfisher bombed through on its way to some unknown destination.

Nearby was the Sanctuary, a thirty-acre nature reserve that Nick Moyes had tirelessly campaigned to create from a landfill site a few years ago. The reserve is situated by the park-and-ride car park in the shadow of the stadium. There's no entry, so you basically view the site either through the perimeter fence or at the specially erected birding blinds that overlook the flooded pits and lightly vegetated mounds. A few weeks into its existence, Nick discovered an obliging Dartford Warbler that stuck around for six weeks. It was the county's first in over 160 years. We managed to connect with at least ten Wheatears, including some very rich buffy-looking birds that certainly seemed to be of the leggier and slightly bigger Greenland race.

Little Ringed Plovers were prevalent, while up in the blue yonder we were treated to several buzzards that flew over during the course of the afternoon. While lying on our backs on the gentle vegetated landfill slope sky watching, we picked up an incredibly high-flying Peregrine plus a possible raptor with slightly angled wings and a small head on a prominent neck. Honey-buzzard? Maybe, though the resultant pictures made its confirmed identity inconclusive. It was exciting nonetheless.

Derby is an unexpectedly good birding venue that deserves a thorough investigation. One of the best ways of enjoying the wildlife the city has to offer is by bike along the River Derwent, as most of the main birding sites are close by. And if you want to become a dead birdwatcher, there's no better place to start.

Eastbourne, Sussex

There's nothing more quintessentially English than sitting on the beach in a deckchair baking in the sun. Gulls yelp and children laugh while you enjoy a Mr Whippy with a lovely flake of chocolate – preferably without the knotted handkerchief on your head. I remember doing just that as a child back in the Seventies when I visited Eastbourne on a day trip with my parents. The memories started flooding back as I surveyed Eastbourne from my lofty lookout on top of Beachy Head on a hot April afternoon. Migrant Swallows buzzed past fresh in off the sea, and my first Common Whitethroats of the season churred from some nearby gorse. Beachy Head is certainly a well-known spot to visit during migration periods and as a result there have been some great birds discovered here. Ring Ouzel, Firecrest and Ortolan Bunting have all been twitched, and a River Warbler spent a day here in 2008. The area is also famous for a variety of other reasons, but I love the fact that the French originally christened the cliffs Beauchef or 'Beautiful Head', which eventually became corrupted to Beachy Head.

I had come to this popular south-coast town to deliver an afternoon talk to the Eastbourne and District Local RSPB Group in Pevensey Bay, just up the coast to the east of Eastbourne, although it is still part of the general district. The beach was literally yards from the hall I was speaking at, so I couldn't resist a quick scan after my talk. My five-minute check resulted in a close-up Sandwich Tern, my first for the year, which was quartering the shoreline looking for morsels in the shimmering sunlight. A couple of the RSPB members who attended my talk kindly offered to give me a quick whistle-stop tour of Eastbourne. How could I refuse an offer like that? On the way into town we passed by Pevensey Levels, a site that I had heard of but never been to. Formerly a large area of marsh, it is now mostly drained but is still the place to find nationally rare plants and invertebrates, including the Fen Raft Spider. Naturally, it is also a good spot for birding and even boasts a 2001 Sociable Plover on its bird list.

There can't be many urban patches in the UK that can claim breeding Bearded Tits. Eastbourne has one in the shape of West Langley Levels, a managed marsh with reed beds that are also occasionally frequented by a couple of Water Buffaloes. They are put to work as part of the habitat management but they are also a massive hit with the kids. The area looks beautiful and as a result is heavily visited by the local populace, who use it for country walks and picnics as well as for angling. The site doesn't appear to be very regularly covered by the local birders but by all accounts it attracts the sort of species that you would expect, with waterfowl and Grey Herons prevalent. On my visit the site was covered with Herring Gulls, and from the reed beds the spluttering and jittering songs of several Reed Warblers rang out. Winter is the time to come to West Langley Levels. It is then that there is the high likelihood of coming across Stonechat, Pintail, Water Rail and Goldeneye, plus occasional scarcer visitors like Smew, Bittern, Mediterranean Gull and, on one occasion, a Caspian Gull.

Perhaps the most urban site I passed through was Hampden Park. It was rammed with people on my visit due to the warm weather. The tennis courts and sports fields are a big draw for the public, as is the large lake that incidentally has a tiny heronry on a small wooded island. It seemed that half of Eastbourne's Herring Gull population had also come to visit, no doubt to capitalise on the free bread that was originally destined for the mouths, or should I say beaks, of the resident Mallards and Canada Geese. The park also contains a large area of woodland that was surprisingly pleasant to walk through. I saw my first Orange-tip Butterfly for the year, and heard Nuthatches calling and Chiffchaffs, Blackcaps, Blackbirds and Chaffinches singing their hearts out. The local council had recently received lottery funding, which it was using to make the locality more nature friendly, projects included silting the main lake to create shorelines and encourage more waterside vegetation.

Hampden Park may not seem the sort of place that a birder would necessarily choose to frequent. This is especially given Eastbourne's positioning between a couple of pretty hotspots, namely Dungeness to the east and Pagham Harbour to the west. It still looked interesting to me. Indeed, although it is not covered with any major regularity by the local birders, Firecrests are apparently regular winter visitors. Birders visiting Eastbourne cannot go too far wrong by taking a peek at this site, especially early in the morning before the crowds descend.

Of course, you may be visiting with the family and be forced to sit in a deckchair with an ice lolly in one hand and a screaming kid in the other, with your birding pass revoked. Not to worry, there is always a spot of seawatching to be had. With so many great birding sites either side of Eastbourne surely something good will fly past. Anything can turn up anywhere at anytime – right?

Exeter, Devon

Can you name a city in the UK where during the course of a year you can scan gathering flocks of Avocets and Brent Geese, hear the songs of Woodlark and Nightjar, and watch Goshawk, Dartford Warbler, Peregrine and Cirl Bunting all within its boundaries? To be honest, until recently I doubted the existence of an urban Shangri-La in the UK. But I was wrong, such a city does exist and it goes by the name of Exeter – the provincial capital of Devon.

When I arrived in Exeter, I decided to drive straight through it and head instead to Dawlish Warren. This renowned coastal birder's paradise on the Exe Estuary is about seven miles south-west of Exeter, sandwiched between a very busy caravan site and a golf course. It's known throughout the land as a winter wader magnet, hosting nationally important gatherings. Plus, it is also well known for its occasional rarity. For instance, did you know that the Warren was the site of Britain's only Great Black-headed Gull in 1859? Cap doffed and respect paid, I headed back to my hotel in Exeter to dream of future firsts for Britain.

The following morning I strolled from the hotel through the city centre to the RSPB's South West Regional Office to meet Tony Whitehead, its public affairs officer and Claire Thomas from Exeter City Council. After a swift cuppa we all walked into town under darkening skies. As we went through the walkway entrance to the Exeter Cathedral square we were serenaded by a migrant Willow Warbler deeply hidden in some bushes. By the time we had located the city's local Peregrine screeching from a ledge on the nearby St Michael's Church, the heavens had opened. We were not the only ones getting a face-full of rain as we peered up at that magnificent falcon. Several members of the public were already staking it out before we arrived, and on seeing the RSPB logo on Tony's jacket they were quick to swamp him with questions.

While he was going through the Spanish Inquisition, I walked down the road and entered the prettily manicured grounds of a retirement home. It was on the crest of a hill so as I looked down the slope I noticed, partially hidden by housing, a well-wooded open space that was unmistakably a cemetery. Claire confirmed that it was St Bartholomew's Cemetery and that it is now a park as the last person was buried here in the 1940s, although parts of it remain consecrated ground. As already mentioned, I have this thing about cemeteries and I try to visit at least one in every city I travel to. Some of the older ones can be great havens for wildlife due to their mature trees and slower pace of life – if you will excuse the pun. But, alas, a shortage of time ruled out my chance of scouting this potentially unbirded sanctuary, as I had to board the Avocet Express.

Well, the Avocet Line to be exact. Having said our goodbyes to Claire, Tony and I boarded a train to Topsham on the Exe Estuary, some three miles south-east of Exeter city centre. The train service does actually carry on until Exmouth, about seven miles from Exeter, taking in a string of Exe Estuary walks and reserves. Unlike the Kite Bus in Gateshead, which is covered with Red Kite insignia and laden inside with tons of kite leaflets and posters, the Avocet Line was a positive let-down with not a single Avocet in sight. Anyway, it was Tony's idea to go green by using public transport and doing the circular tour. His idea was to take in the birds at the RSPB's Bowling Green Marsh Reserve, bordered on the east by the estuary of the River Clyst and to the north by the Avocet Line, then end back at the station.

Thus, after a short walk from Topsham Station we were soon ensconced in a hide overlooking the marsh and studying an Osprey that had chosen to linger there for the past few days. Despite it being a fairly glum day, it was clear that it was an attractive site. We registered great views of Bar-tailed Godwits, my first returning hirundines of the spring and a glorious winter-plumaged Little Gull. The star bird for me, though, was the solitary White Wagtail that I discovered walking up a channel. Its black bib stood out like a shield against its pallid grey back. Tony explained that winter produces some of the best birding spectacles, with large flocks of wintering waders and waterfowl including hundreds of Avocets.

Leaving the hide and continuing our walk, Tony showed me the small roadside pool where Bowling Green's famous long-staying Glossy Ibis gave close-up views to its adoring audience. We made our way to the raised, exposed viewing platform further down the lane that offered a great vista of the mouth of the River Clyst joining the River Exe Estuary. The wind with spitting rain was whipping up something serious, and only a few Curlew and a couple of hardy Little Egrets were on show, although a large flock of around 120 Black-tailed Godwits with a solitary Bar-tail did wisp through as we were leaving the platform. Tony enthused about the Brent Geese, mergansers and Avocets that can be seen on the River Exe during the winter, as we headed back towards Topsham.

Exeter was certainly an eye-opener. Sitting in a quaint Topsham teashop I came to the conclusion that I had barely scratched the surface as I tucked into my scone that was so lavishly filled with cream and jam. I decided that I needed to come and ponder in Exeter's teashops much more often.

Gibraltar Point, Lincolnshire

It's great to wake up and go birding bathed in sunshine after a previously dreadful day of weather. I've had a few of those memorable days. But on the flip side, have you ever felt singled out by weather systems that seem to plague you every time you pick up a pair of binoculars? Recently, I would just have to don a pair of bins and head for the door and a shroud of fog would descend upon proceedings, resulting in me chasing murky shadows. It once happened to me while I was in the middle of a session at the Scrubs, and I have experienced it to a more minor degree while birding in Northumberland. A while ago I boarded an evening train to Skegness and the words of that immortal record by Crowded House, 'Weather with You', rang in my ears as an ominous ground-hugging mist gathered again.

My plan was to wake up in Skeggy and indulge in some not-so-urban birding at Gibraltar Point Nature Reserve under the watchful guidance of site manager Kevin Wilson. The following morning, excited by the prospect, I peered out of my hotel window, only to be greeted by a thick wall of fog. Great. Maybe it will burn off a bit later, I thought. As Kevin and I drove to the reserve through the posher parts of Skegness that bordered the Seacroft Golf Course, which is a site of special scientific interest and part of the Gibraltar Point complex, I had a terrible feeling that the fog that had been following me around was here to stay.

Gib Point is one of those special places that I had always wanted to visit and it was brilliant to get a personal invitation from the nature reserve staff. Managed by the Lincolnshire Wildlife Trust, the reserve stretches from the southern edge of Skegness into the mouth of the Wash. It covers some 1,000 acres and encompasses a range of habitats that include freshwater marshes, grassland and the most extensive sand dunes and saltmarsh on the Lincolnshire coastline. Over the years Gib Point has played host to a number of fantastic birds like Thrush Nightingale. But perhaps the site's biggest claim to fame was the American Redstart found there in 1982. Apparently, at one stage it stood in close proximity to an Isabelline Shrike. How crazy was that? The place has also received both of Lincolnshire's only records of Terek

Sandpiper, and what about throwing a Sora Rail and Greater Yellowlegs into the mix? I could not understand how this great site could be currently underwatched when it was as clear as day that there was so much potential for falls of interesting migrants.

Speaking about things being as clear as day, we could barely see a thing when we stopped off at the ringing hut near the reserve centre to check up on the activity. We entered the hut to see the ringers in the closing stages of processing a Robin they had just extracted from the mist nets. The next bird that they pulled out of the bag was rather more thrilling: an impressive-looking, if slightly underweight, Woodcock. They are such amazing birds when seen at close quarters; their cryptic markings are truly dazzling. When it was finally released the bird duly flapped towards a nearby hollow where it sat and pretended that it was invisible.

Standing outside the visitor centre was a gloomy affair with visibility down to perhaps twenty metres, and the pea souper was still showing no signs of letting up. A male Black Redstart hopping around the building did its best to cheer me up as a group of birders that Kevin was due to take on a guided walk accumulated around us. Undeterred by the poor visibility, Kevin began his tour around the reserve by instructing us to switch on our ears. It is a funny sensation walking around birding by ear as my strengths lie firmly in the visual medium. I have a heightened sense of movement that often enables me to notice distant birds darting around. However, when it comes to hearing birds I'm rubbish, largely due to the profound deafness in my left ear. Ask me to pinpoint a warbler calling from a section of woodland, for instance, and I would have to either concede defeat or deploy 'the Force' and basically guess where the sounds were emanating from.

No, I wasn't looking forward to birding in the fog. Fortunately, Kevin had bionic ears and was quickly picking out unseen Curlews, Redshanks and distant Pink-footed

Geese. I redeemed a little of my self-esteem by noticing a male Reed Bunting in the cloudy haze on top of a nearby bush. Later, while we were in the Plantation area watching some foraging Goldcrests in a tit flock with a couple of calling Chiffchaffs, I also became aware of a solitary Spoonbill as it silently circled low overhead before disappearing behind the trees. That was a slightly strange experience because the silhouette of the Spoonbill seemed eerie and totally unfamiliar as it cut through the fog. The Spoonie was the last bird we saw because thereafter the fog got denser and denser. Quite frankly, I could have been anywhere in the UK as it all looked the same – grey.

So that was my Gibraltar Point experience. My verdict? I didn't see any of it but it sounded good. I'll have to come back again for a better look.

Glasgow, Lanarkshire

I decided to journey north to Scotland with my statement on Facebook exclaiming that I was going to do some kissing in Glasgow and mulling on Mull. Well, I didn't indulge in kissing of any kind while in Glasgow, but I did manage to get reacquainted with a city whose streets I hadn't stalked for over twenty years. As usual, I timed my visit impeccably – during the coldest snap in thirty years. Snow and ice abounded.

Glaswegians are very strong-minded, passionate people and not just about football. Iain Gibson and Jim Coyle, my guides from Glasgow City Council, were typical Glaswegians: totally captivating me with their infectious enthusiasm for the city's wildlife. Their fervour was well justified, as Glasgow is certainly an urban gem. The city has long been known as the 'Dear Green Place' and, boy, there were a lot of green spaces. There are seven designated Local Nature Reserves across the metropolis that support a range of habitats including wetlands and raised bogs covering a total area of 618 acres.

On my arrival I was whisked away to Pollok Country Park – 173 acres of parkland and woodland, one of Scotland's top tourist hotspots and home to the largest Scottish Magpie roost, which peaked at 538 birds in February 1998. Cold-looking Fieldfares roved over the snowy terrain as I relaxed, drinking tea, in the warmth of the Burrell Collection cafe. If you come to the park during the summer months you will soon discover a good range of the typical woodland species, including healthy numbers of the common warblers along with the city's only breeding pair of Nuthatches. We took a quick walk to the feeding station situated behind the old stable courtyard near Pollok House to twitch the Nuthatches coming to the feeders, but instead we were treated to the antics of the numerous tits and Chaffinches.

I liked the look of the park and on asking about the more unusual sightings I was told that on one occasion a migrant Ring Ouzel was seen feeding on Yew berries along the riverside walk.

We then headed to the north-east side of the city to Millichen Farms (pronounced Millikan). It is privately owned land that currently has limited parking facilities, though the site can be viewed from Millichen Road. The area supports a rich farmland bird community with often spectacular flocks of geese, other wildfowl and waders during the autumn and winter. Judging from the photographs, the geese flocks are something to behold, with internationally important gatherings of the Icelandic race of Greylag. In recent winters up to 1,600 Pink-feet have gathered, as well as lesser numbers of Canada and Barnacle Geese. Of course, the day I decided to rock up the area looked like a winter wonderland with not a goose in sight.

However, we did manage to find a few Teal, at least two Water Rails, as many as twenty Snipe and, best of all, a couple of Jack Snipe, all feeding in a small unfrozen burn (stream). Nearby, flocks of both sparrow species and a sprinkling of Yellowhammers noisily filled the adjacent hedgerows and shrubs. Up to thirteen pairs of Tree Sparrows breed here, and as they are a nationally decreasing species it was good to hear that there was a programme of hedgerow management, nestbox provision and winter supplementary feeding in place specifically for them. I also learnt that Bramblings and Twite occasionally flock in the area.

Iain and Jim explained that autumn is the time to observe the waders that use this site. Aside from regulars like Curlew and Lapwing you could uncover goodies like Little Stint, Ruff and Spotted Redshank, and there has been a record of Pectoral Sandpiper. It's also a good spot for gulls, with Ring-billed, Glaucous, Iceland and Kumlien's Gulls all on the list. The guys informed me that there are plans afoot to make the area into a wetland reserve. I'm sure that it would make an amazing urban site.

By now the day was beginning to draw in, so we decided to move on to our last port of call, the famous Hogganfield Park Local Nature Reserve. Its fame emanates from the apparently regular winter occurrences of its legendary speciality bird: the Jack Snipe. The nature reserve features a large loch with a wooded island, managed areas of grassland, scrub woodland and a fairly recently created extensive marsh with a pond.

More than seventy-five per cent of the loch was frozen over on my visit and all the teeming waterfowl were packed in a small, ice-free stretch of water by the car park. There was a feeding frenzy in progress as members of the public were throwing bread to the birds. At first glance the mêlée seemed to consist of Black-headed Gulls, Coots, Mute Swans, Greylags and Tufted Ducks, but the guys were quick to point out the presence of wild Goosanders, Goldeneyes and Whooper Swans among the throng squabbling over the bread – often within feet of us. The guys explained that Goosanders coming to bread was a recent phenomenon, which made me think that perhaps not all non-Mallards coming to bread are escapes.

Of course, Hogganfield is a great place to visit at any time of the year and has a very impressive list that includes some interesting migrants. I left Glasgow very enlightened. It's an absolute urban haven with plenty of surprises waiting to be discovered.

Hartlepool, County Durham

I would like to make an announcement: Hartlepool is an urban birding nirvana.

Forget all the negative press that you may have heard about this small northern town, and discard from your mind the notion that the birding is not worth coming here for. Some of you may well be wise to the fact that a good many decent birds have turned up on Hartlepool Headland, not least 2011's famous White-throated Robin and a more recent much-sweated-over Western Orphean Warbler. But how many people know that the whole of the headland is a great place to wander around in pursuit of birds at almost any time of the year?

When I showed up on an overcast morning during late October, it was a classic case of 'you should have been here three days ago…' Earlier that week, while the east coast of England was being coated with a thick layer of fog and an even thicker sprinkling of lost migrants, Hartlepool Headland was also enjoying a fantastic fall. Among the more expected

Redwings, Fieldfares and Goldcrests were hoards of Robins with Bramblings and a few scarcer migrants like Yellow-browed Warblers.

Even though I arrived after the event there was still loads of movement, with small groups of Starlings steadily coming in off the sea and roving flocks of winter thrushes streaming over the rooftops. The bushes were not dripping with migrants, as previously, but there were still plenty of foraging Goldcrests within the miscellaneous tit flocks to enjoy. One of the main epicentres of the much-vaunted mass avian movements was at the Bowling Green, an innocuous-looking square of short mown grass lined by shrubs and bushes occasionally interspersed by small cut rides. The rides were for mist nets, run by ringer Chris Brown, my host, who was showing me around the town, along with Toby Collett, assistant warden at Saltholme RSPB Reserve. Despite its ordinary appearance, the Bowling Green can be a magical place for picking up migrants and history has proved this. You could bowl up – to excuse the pun – and see absolutely nothing, and come back in an hour to be confronted by a Pallas's Warbler. As if to illustrate this, Chris pointed to a ride and explained that two Paddyfield Warblers had been trapped there over the years.

There was another bowling green across the road. It was the site where the White-throated Robin was released after being ringed. There is an amusing story about locals playing bowls while having their bowling techniques vociferously analysed by crowds of twitchers seeking a glimpse of the robin. Another major migrant trap are the council allotments, the only ones on the headland and yet another site that looked totally underwhelming at first glance. Despite being working allotments, they were in reality a tiny plot of land with a few onions growing and patches of scrubby cover. It was on this hallowed turf that the White-throated Robin was originally discovered. As is my custom, I immediately sunk to my knees to pay my respects. Chris has mist-netted all sorts of migrants in there including Blackcaps, Black Redstarts, a Bluethroat and a Red-breasted Flycatcher. If you ever get the chance to visit these allotments it would be rude not to stand on the bench that overlooks the wall to the Doctor's Garden. This famous garden is where the White-throated Robin spent the bulk of its time, and it is the wall on the street side of the garden that twitchers famously propped up their ladders on, and that is next to where they stood on tops of vans to view the bird, as posted on YouTube.

Over the years the Doctor's Garden has been a veritable treasure trove for rarities. Wanderers like Little Bunting, Subalpine Warbler and Thrush Nightingale have all been watched enjoying the haven that the garden provides. The Doctor's lawn, which isn't that extensive, has hosted legendary flocks of hundreds of Redwings and Fieldfares – not over a period of time but on the same day at the same time. They must have all been stacked on top of each other.

After visiting the aforementioned birding shrines we took to the streets, quite literally, and began to walk around peering into gardens and even searching the exteriors of the buildings for signs of migrants. During a good fall birds will perch and take cover anywhere whether natural or man-made, and the guys fondly recalled days when stacks of thrushes lined rooftops and gutters.

Practically anywhere on the headland is fair game for interesting birding and I have not even mentioned the seawatching that can be pretty fruitful too. A quick thirty-minute scan resulted in plenty of Common Scoters offshore, plus I saw at least twenty Little Auks, my first for the year. My favourite area was the Jewish Cemetery and scrub at the northern end of the headland. Aside from a small area of headstones that at the time was holding a visiting male Black Redstart, there was also the adjoining patch of scrubland that attracts Long-eared Owls and assorted migrants. The potential here seemed enormous and I could have very easily spent half the day just sifting through the bushes. I came to the conclusion that Hartlepool is a truly exciting place and one that I will have to visit again in the hope of striking it lucky. Maybe I should buy a house here…

Hull, Yorkshire

I've been to Hull and back and I'll tell you what, it was actually all right. Despite being steeped in history, Kingston upon Hull, to give the city its full name (or King's town upon Hull of yore, if you were birding in the time of King Edward I), has a fairly negative reputation as an ornithological venue. It's one of those classic cases of if you don't come from there what are you doing going there; surely there are no birds? Being a sucker for a good challenge, it was just the encouragement I needed. However, to be honest I was questioning myself as I journeyed up on the train during a warm summer's night, and continued to doubt myself when I visited East Park the following day, smack in the middle of town.

When I put out the message on Twitter asking for any citizens of Hull, former or otherwise, to come forward and suggest birding sites, East Park was mentioned several times. I found it to be a fairly bland municipal park with a small animal enclosure and a large elongated lake within its grounds. But apparently it is the lake that attracts the most interesting birds, with a good selection of wintering wildfowl that sometimes includes Goosander. All I could muster were hordes of marauding teenagers, feral Greylag Geese, a few Mallards and a couple of Peacocks sitting on a low fence. Add to that the vision of a big, ugly grey cloud hanging ominously overhead and the scene was set. My heart was sinking.

Then came Les Johnson of the Hull RSPB Group. He had agreed to guide me through the murk like a beacon at a misty headland to happier avian hunting grounds. Soon, after a short car ride I was watching a flock of Tree Sparrows sheltering in trees alongside a field in driving wind and rain. I was experiencing the joys of an English summer at Paull Holme Strays, approximately 200 acres of relatively new intertidal habitat on the banks of the mighty Humber created by the Environment Agency as part of a flood-risk-management scheme. The tide was drifting out and the waders were drifting in with loads of Redshank, Curlew and a couple of Green Sandpipers heading the cast. This reserve is also well known locally for the flocks of Avocet that sometimes congregate. Distantly, I noticed two mobbed Marsh Harriers dipping into a reed bed. All of this within sight of civilisation – that was more like it.

Paull Holme Strays had hosted a transatlantic Lesser Yellowlegs a few days before my visit, and during the winter draws in more waders, Hen Harriers and coasting Short-eared Owls. The Humber itself is a fascinating birding spot and if you ever come to the docks to catch a ferry to Zeebrugge, spare some time to watch its waters because it has recently been discovered that it is part of a flyway for a multitude of seabirds, including skuas, terns and even shearwaters.

Les escorted me to a few more sites that were on the outskirts of the city, including Welton Waters, three scrub and reed-fringed lakes that are also beloved by the local boating fraternity. Brough Haven, situated a couple of miles outside Hull itself, was another attractive area alongside the Humber. We were able to pull up and view the exposed mud as the rain spat. A solitary, steely Grey Heron was our only reward.

Perhaps the most attractive place Les showed me was North Cave Wetlands, situated near Hotham and run by the Yorkshire Wildlife Trust. By this point I realised that we were sufficiently far away from Hull to start to seriously stretch the definition of urban birding. However, after a tasty, stomach-warming bacon buttie at the Wild Bird snack van by the entrance to the reserve, a stroll around the site was called for. Despite the rain we saw some good stuff, including some corking views of a couple of close-up Green Sandpipers.

The following day I decided to be a bit naughty (well, indulgent) and made the thirty-mile trip to Spurn Head. I had never been before and my rationale was that I could not be in Hull and not visit Spurn. I ended up spending time chatting with the warden, Andy Gibson, himself a Hull resident. It was time well spent because he told me about many more sites in the city that I was not aware of, the sorts of places that only a local would know about.

He directed me to Priory Fields, situated between Hull and the practically adjoining town of Cottingham. It is an area of grassland that is home to a range of the common warbler and other passerine species. Then there was Bransholme Sewage Works next to the River Hull and a locally good place for breeding Reed Buntings. Sculcoates Cemetery fulfilled my desire to find a cemetery wherever I go. It also satisfies the needs and desires of Bullfinches, some of the common summering warblers and patrolling Sparrowhawks.

So as you can see, appearances can be very deceptive as Hull not only has some interesting inner-city sites but it is centrally positioned for some nationally important birding Meccas.

Leicester, Leicestershire

John Hague – birder, twitcher, comedian and adopted son of Leicester told me about Harold the Whooper Swan while

I stood freezing off my proverbials in the middle of a flooded meadow. It was supposed to be spring and I thanked God that I hadn't gone into a full body moult and dispensed with my winter fleece because despite the blue sky it was absolutely brass monkeys.

We were visiting Cossington Meadows on the outskirts of Leicester, a seventy-acre Leicester and Rutland Wildlife Trust run nature reserve on the floodplain of the River Soar. In common with practically everywhere else, this area of flooded gravel and sand extraction pits had had more than its fair share of rain and the water levels had risen very high. I was surprised to see small flocks of Wigeon still nervously loitering on the lakes in the company of some even edgier Teal. I began scanning skywards for the first Swallows of the spring but the gloriously blue skies were empty aside from traversing Cormorants, Grey Herons and ducks including Mallard, Gadwall and the occasional Shoveler. I did, however, manage to find a migrant Wheatear in a field.

This really interesting patch regularly attracts those supreme aerial warriors, Hobbies and Peregrines, and held four Short-eared Owls in a recent winter. Yellow Wagtails are expected migrants and there's a good winter Reed Bunting roost here too. It is also one of the best places in Leicester to find a Cuckoo. Cossington Meadows has a decent rarity track record, including its famous Black-winged Pratincole during the Seventies and, more recently, a stonking Little Bittern.

Leicester is certainly one of those places where you just would not expect to find an abundance of birds, but even I had to raise an eyebrow at the array of sites that could reward vigilant birders. One thing that struck me was the amount of brownfield sites in and around the city. Due to the recession, many of them had not been built upon and the recent excessive rain had made some become mini-wetlands. We stopped at a few and I was surprised at the number of Little Ringed Plovers that had taken up residence. The industrial estate at Rothley Lodge held LRPs, Lapwings and a solitary urban Golden Plover, while the wasteland at Grove Park had obligatory Little Ringed Plovers running around, foraging Meadow Pipits and Pied Wagtails, plus a couple of Snipe. John told me that a few days previously this particular site had a count of twenty Snipe and several Jack Snipe. It will be a shame if once the economy picks up these temporary oases will be no more.

We bumped into a couple more LRPs on some of the derelict land on the Raynesway Estate where John had found Wheatears in previous springs. Nearby at Watermead Country Park we visited the locally dubbed Pec Pits to pay homage to the place where a couple of Pectoral Sandpipers chose to hang out one year. We then moved on to view a very waterlogged Wanlip Meadow, a site that regularly gets Temminck's Stint and was also home to the Pec Pits' Pectoral Sandpipers. It was here that I finally started seeing the hirundines I had desired, flashing overhead.

Of course, being the Urban Birder, I needed John to show me what the city centre had to offer me. He rose to the challenge and was soon escorting me around Victoria Park, a municipal open space with a small area of wet woodland that was sometimes a migrant trap. It was the sort of place that with regular checks could well reward watchful birders. Evington Park was another unlikely spot with an open-air gym and manicured gardens that sported a couple of small ponds each housing populations of both Great Crested and Smooth Newts. A Yellow-browed Warbler once turned up in a hedge near the newts and a Pied Flycatcher was also a recent find. Welford Road Cemetery satisfied my curiosity for urban burial places and also showed signs of being a place deserving of more regular surveillance, as there were some good scrubby areas around the graves that were less well tended.

Our urban day ended over an alfresco cup of peppermint tea in the King's Lock Tea Rooms overlooking the flooded meadows of the similarly named nature reserve. Our quest was Ring Ouzels, as this was a regular spot. Instead, we got Woodpigeons and our only excitement was a piece of debris being blown down the river that looked not unlike the fin of a miniature Great White Shark.

Harold is a Whooper Swan that originally came to the Leicester area as an immature bird around three winters ago, promptly disappearing to reappear as an adult bearing a ring from Cumbria and has stayed ever since. Despite searching for him at several of his preferred haunts we never caught sight of him.

Lowestoft, Suffolk

The Stone Roses have just done it, Blur did it and even Duran Duran managed it, but would it happen to me? Could the old partnership between my erstwhile main birding buddy Cornelius Ravenwing III and I be rekindled? We were a birding duo who chased feathered dreams through adolescence into young adulthood. The plan was to burst back onto the scene in our chosen venue of Lowestoft on the Suffolk coast, and fresh after finding an autumnal mega-rarity there, we would chase around the nation as we once did in search of birds. However, that great comeback stalled at the first hurdle. At the last minute Cornelius had decided to spend the morning in bed with a new girlfriend instead of shaking bushes with me among the coastal scrub.

I'll teach him for reneging on arrangements, I thought, as I crept around looking for tired migrants in the company of

Lowestoft local birders Andrew Easton and Steve Jones. We had met along the coast at Links Road car park. This was no normal car park as it had recently boasted a Woodchat Shrike that stayed for a week. A cursory check of the gulls that were loafing by a puddle revealed no rarities, just a group of twelve Black-headed Gulls. I noticed that one of them was visibly taller than others, standing on longer flesh-coloured legs, quite different from the reddish pins that its congeners were sporting. I toyed with the idea of it being a winter-plumaged Slender-billed Gull. It too has flesh-coloured legs, but after consulting with the guys we had to reluctantly agree that it was just an aberrant Black-headed. It would have been a great find to taunt my distracted friend with but maybe it was a precursor for what was to come?

I was being shown the interesting birding venues along the northern edge of this totally underwatched town. Andrew, who had been birding in Lowestoft since the mid-Seventies, explained how the area and its environs was viewed as the poor cousin to nearby Minsmere and to the Mecca that is Norfolk. Whenever a rarity was found elsewhere in East Anglia the Lowestoft birders would rush off to twitch it instead of being inspired to search their own local doorstep. Had they done that far more special birds would have been uncovered.

I certainly began to see how attractive the habitat was for birds when we entered Warren House Wood across the road and north of the car park. It had rarities written all over it. Slowly walking through the tangled paths we came across small tit flocks captained by Long-tailed Tits, while small squadrons of Siskins peppered with the occasional Skylark passed south overhead. The wood's main claim to fame was Suffolk's first Red-eyed Vireo back in 1988. It turned out to be the first of four to be seen in the Lowestoft area over the years. We eventually surfaced from the wood into Gunton Warren, an area of scrub and bracken that butted up to the beach. The Warren can attract a small contingent of wintering

Dartford Warblers but it is also a very popular place for dog walkers, so an early-morning visit to catch migrants before they moved on was a good idea. Our rummage around resulted in a few wary Blackbirds, several Meadow Pipits and lots of fantasising about finding a wayward Rufous-tailed Robin.

We drifted back south past the car park along the sea wall. Eventually we reached Flycatcher Alley, so named, as you may have guessed, by birders back in 1910 impressed with its preponderance of flycatchers. We didn't see anything for ages until a party of Crossbills flew over to excite us, followed by a very brown-and-white Chiffchaff that popped out of the herbage for us to inspect. It looked pretty good for a 'Siberian' Chiffchaff, a scarce visitor to these shores. Andrew then caused me to heavily salivate when he recounted the species that had been found earlier in the autumn: Icterine, Cetti's and at least three Yellow-browed Warblers.

Further south was Sparrows Nest Park – a quite wooded and very prettily landscaped open space with an open-air theatre, bowling green and the like. The park was screaming migrant trap to me and I was not wrong. Its roll-call of great birds included a recent Olive-backed Pipit, one of the famous Red-eyed Vireos, a Red-flanked Bluetail and a male Collared Flycatcher, among others. The magnet drawing them in, as well as the more common migrants, was a still functional lighthouse in the south-west corner throwing out its inviting light to the passing avian travellers. We managed to connect with the first Brambling of the autumn as well as a brief Firecrest with a roving tit flock. This spot was well worth a visit and it conveniently had a cafe on site to enable us to indulge in a celebratory cup of tea.

I spent the end of a very interesting day seawatching at Ness Point, the most eastern point of Britain. In days of old there used to be a sewage outfall here that attracted interesting gulls, including regular Sabine's Gulls and mammoth flocks of up to forty-two Purple Sandpipers on the rocks. Along with Andrew, Steve and several friendly local birders I managed to see a Red-throated Diver and at least five Great Skuas out to sea chasing anything with wings. Nearby, a female Black Redstart was performing, and on the rocky shore a lone Purple Sandpiper was feeding. So, Cornelius, who had the better day? Surely there's no contest.

Manchester, Lancashire

If you were to ask me what English city I would choose to live in aside from London, then I would have to answer, Manchester. I have a connection with this city that stretches back to my teens, when I first used to come up to watch Manchester United.

In those days my Mancunian urban birding was during the football season and largely confined to the Salford Quays area on the doorstep of Old Trafford, Manchester United's home ground. Over the years I have watched many great games embellished by overflying gulls, Cormorants and the occasional marauding Sparrowhawk. But sometimes the lure of rare birds was too strong and, on the odd occasion, I would miss a match to twitch a lost waif like the Black-faced Bunting that turned up at nearby Pennington Flash back in 1994.

This time, though, I had come to my second favourite British city purely to watch the birds that this conurbation has to offer. I arrived in Manchester Piccadilly Station under a wet Saturday-morning sky, knackered. I had literally had two minutes' sleep before getting the first train out of

London at an ungodly hour after doing a late-night radio talk show in the capital.

I picked up a hire car and headed north-west towards Irlam to visit an area collectively known as Chat Moss. On the way, while passing through Eccles, I happened across Peel Green Cemetery. I couldn't resist. I had to stop and take a look. The cemetery was a classic post-1880s burial ground with some thirty-two acres of mixed woodland. As I strolled around grey clouds evaporated and morphed into a hot and sunny day – typical British weather. I sat on a bench overlooking a richly vegetated graveyard, bordered by stands of oaks and other mature trees, enjoying the cacophony of avian sound. Multitudes of Goldfinches twittered from the treetops around me as Great and Blue Tits went about feeding their fledglings, and Blackcaps and a distant White-throat sang. The mixture of fatigue, morning warmth and the tranquil birdsong quickly had my eyelids drooping, and soon my snoring became part of the morning chorus.

Later that morning I arrived at the Chat Moss area in Irlam, temporarily rested and looking forward to my next bout of urban exploration. When describing Irlam I have to use the term urban loosely, as many of the areas that I drove through looked distinctly rural. It was a perfect vision of how much of Manchester must have been like back in the day. However, Manchester city skyline on the horizon reminded me of just how close I was to civilisation.

Chat Moss is the collective term for a large area of peat bog that makes up thirty per cent of the city of Salford. It's a 10.6-square-mile green space that largely consists of arable farmland, vegetable fields, virgin mosslands and the county's largest wood, Botany Bay Wood. Some of the fields sported healthy-looking hedgerows and shrubby drainage ditches. I encountered the typical country birds that you would expect, with Yellowhammers announcing themselves from the tops of bushes, Reed Buntings monotonously singing, Swallows swooping around and mighty Common Buzzards leisurely winging their way over the countryside. Apparently, Quail also sing here most years. At one point I pulled over by some farm buildings to examine a flock of small birds that I had noticed on a wall and was rewarded with the sight of my first Tree Sparrows of the year. I was in my element.

I could have spent the day in this area alone but time dictated that I head back east towards the city. So I decided to swing by Chorlton Water Park, the site of my first-ever Ferruginous Duck back in the mid-Nineties. This reserve is a lake that was constructed during the Seventies and has since become a nationally important area for waterfowl, and indeed a lot of unusual species have turned up here. There were no such notables to be found on my visit because the heavens had decided to reopen as I arrived in the car park. There was something quite hypnotic about peering through a car windscreen at lashing rain in between swishing wipers. Weariness soon started to set in again, so back to the city I headed to return the car and find a hotel to get an early night.

The next morning I awoke to a mixture of cloud, rain and blazing sunshine – sound familiar? I dragged myself from my comfy bed and after breakfast, I walked over to the nearby bustling market square to visit the RSPB's Peregrine Watchpoint. The staff quickly pointed out one of the four fledged youngsters sitting on the 'E' of the 'ARNDALE' lettering on the side of the Arndale Centre building.

With my train departure time back to London rapidly approaching, I walked to a quiet backstreet close to the watchpoint to stare up at some old buildings. After a few minutes I was watching a pair of Grey Wagtails bringing food to a nest hole near a window ledge. While marvelling at this riparian delight raising its family in the heart of a city miles from the nearest raging river, the second special tenant of the building made an appearance. A female Black Redstart popped out of her nesting site several windows along from her long-tailed neighbours.

Thus ended a weekend of rain, sun, sleep and birds.

Milton Keynes, Buckinghamshire

It was a dark, rainy night in the mid-Eighties. I remember it well. I'm crammed in the back seat of a car between two lardy mates driving aimlessly around Milton Keynes looking for an address in the pouring rain. It was in the days before satnavs, and thoughtlessly we didn't have a street map as well. We were onto a loser. The car chugged along, its boot laden with heavy record boxes. We circled around innumerable roundabouts looking for the house in which I'd be DJing that night. We found the premises eventually. The party was rubbish and I vowed never to return. That was my overriding memory of Milton Keynes – dark, wet, full of roundabouts and majorly disappointing. Cut. Fade to black.

It's August 2010, the sun is shining and again I'm motoring up the M1 towards Milton Keynes, though this time I have modern technology at my disposal. I was lured back at the invitation of local birders Mark and Gill Baker who I had met at Slimbridge a few months previously. They had tempted me by promising to show me a side to MK that I certainly did not appreciate in the past. I always thought that this Buckinghamshire town was completely newly built to accommodate the London overflow. Parts of it are and have been fashioned very much like a US city with a grid system of roads and plenty of roundabouts, but it also incorporates a number of medieval villages. And I was astounded by how green the place was. The town planners had obviously designed it with that in mind. Apparently, there are four trees to every person. Wherever I went there were cycle paths, parks, green corridors and lakes – loads of lakes. As my host Mark proudly proclaimed, Milton Keynes is the city of lakes. The town's thirty-four square miles also have large areas of housing, commercial properties and one of Europe's biggest shopping centres, meaning that balancing lakes had to be created to deal with the excess run-off.

Perhaps the most nationally famous of those water bodies is Willen Lake, long known for its ability to attract interesting gulls, Black Terns and waders, including a recent Wilson's Phalarope. Split by the A509, the southern portion of the lake was like hell on Earth. It was packed with the tattooed denizens of the town pushing buggies, walking mastiffs and enjoying the very popular watersports facilities. A short walk to the northern portion of the lake seemed as though we had crossed an invisible barrier into an area much less populated by day-trippers. We sat in a hide and enjoyed a couple of Greenshanks and a female Red-crested Pochard among the more numerous Cormorants and Mallards. Plunge diving nearby were some family groups of Common Terns that were sharing the same stretch of water as families of Great Crested Grebes. How tranquil.

But there was more to meet the curious eyes of a birder than that hotspot. Perhaps the most interesting place we visited was a relatively unknown one, the Hanson Environmental Study Centre, north-west of Willen Lake. It's a ninety-acre area of meadows, woodland, reed beds, ponds and, of course, a large lake. The Milton Keynes Council, to whom you have to apply for a membership card and a key in order to gain access, manages the site. It is Mark and Gill's local patch and boasts breeding Grasshopper Warblers, Nightingales and Bullfinches, along with nineteen species of dragonfly. Both Barn Owl and Little Owl occur, and while walking around and watching the plentiful Willow Warblers I got to grips with several Marsh Tits. From the main hide overlooking the lake at least six Little Egrets could be seen roosting in a waterside tree, we heard a Kingfisher calling and, to crown the occasion, a Hobby chose to hunt insects high over the lake. I could see why Mark and Gill loved this site, as it was a patch little known even to locals.

MK's combination of woodland, riverside parks and numerous lakes provides a contiguous tree-lined green corridor that dissects the town. It is possible to cycle throughout this network along the Redway, stopping at your leisure to take in the avian sights. I roamed through terrain that yielded a Little Owl and seemed perfect for passage Wheatears. Exploring the woodland to the south and east would have been a good option too. There's even birding to be done in central Milton Keynes around Campbell Park by the huge shopping centre. We grabbed a cup of tea in the complex, then strolled out into the adjoining park to enjoy the vista of the surrounding countryside and Willen Lake, largely hidden from view by some woodland that potentially harbours both crests and Crossbills.

The park is the highest point in MK and the grassy slopes of the hill I was standing on were favourite haunts for thrushes during the winter. The previous year there were exceptional numbers of Redwings and Fieldfares. Despite the warm weather and the presence of thermals, enjoyed by rafts of gliders circling overhead, we did not see any of the expected Red Kites and Common Buzzards riding on the warm air. But that didn't matter because my head had been well and truly turned by what I encountered in Milton Keynes.

Northumberland and North Tyneside

Pre-dawn darkness and icy spitting rain enveloped me as I emerged from my hotel just outside Newcastle Airport to meet the Northumberland and Tyneside Bird Club's honorary secretary, Alan Tilmouth. He had agreed to my last-minute request to be shown around some of the urban birding hotspots along the coastal areas of Northumberland and North Tyneside. This was a whole swathe of England about which I had nothing but a cursory knowledge. In fact, the last time I raised a pair of binoculars in Northumberland was back in the early Nineties when I discovered a party of Whooper Swans on some unnamed roadside lake while I was driving back from business meetings in Scotland. I badly needed to update my Northumberland birding.

Thus began a very rapid whistle-stop tour with the aim of arriving at Washington WWT just south of Newcastle in North Tyneside by the afternoon. Alan duly took me to Blyth South Harbour, the place where it all began for him as a birder. As a kid he used to trawl the area for birds and delighted in finding Skemies – feral pigeons to you and I – that nested beneath his feet in the wooden rafters of the harbour. Today, it is still a working port; however, in common with many ports in the British Isles it used to be a lot busier. The place looked fascinating to me, with small areas of beach sandwiched between the frequent groynes. We spent a bit of time scoping the assorted gulls on the shoreline and unsuccessfully searching for some reported Snow Buntings in the blustery grey light. There were also a number of large storage sheds that are used as roosting sites by significant numbers of gulls, plus waders like Ringed Plovers and Dunlins. The harbour itself also had a good population of Eiders, a few of which made their presence known as I began to scan with my bins.

I was just getting my eye in at this interesting patch when we had to move on. Prestwick Carr was calling. Four miles north of the centre of Newcastle, it is one of the few remaining lowland peat bogs in the UK. Back in the day, in the 19th century, it boasted breeding Corncrakes, Ruff and Bitterns. However, over the years it got drained and dyked for farming, resulting in the loss of those spectacular breeding species. Recently, the Northumberland Wildlife Trust instigated an ongoing project to reflood some of the area in an attempt to restore its original state. It has already attracted non-breeding Ruff, along with Redshank and good numbers of wintering waterfowl. As Alan was speaking I noticed some movement in a nearby hedgerow, which turned out to be one of Prestwick Carr's more endearing inhabitants: a Willow Tit. I was delighted, as I could not remember the last time I had seen one of those sprites. Sections of the hedgerows were also alive with roving Fieldfares and Redwings, but we dipped out on the Great Grey Shrike that had been frequenting the area for the previous few days. I blamed the poor light.

Big Water Country Park was certainly an urban area. It is a big, round lake circuited by parkland and rimmed by urbanity. There is a thriving Tree Sparrow colony nearby and the lake was teeming with Common and Black-headed Gulls, with a good crowd of Wigeon present too. Lurking among them was a solitary, continually diving female Lesser Scaup, my first in Britain. The country park was no stranger to rarities because it has hosted a singing Great Reed Warbler and a famous female Pine Bunting in 1991. Now in full-on twitching mode, we thought it would be rude to bypass a Greater Yellowlegs that had been hanging out at Hauxley. That was yet another UK tick for me and the attendant Grey Phalarope made it even sweeter.

Perhaps the most surprising place I visited was Gosforth Park Nature Reserve right on Newcastle's doorstep. Walking into this 145-acre reserve was like stepping into Narnia – the profusion of wildlife was incredible given its close proximity to the city. It has the only urban population of Red Squirrels in England, several roosting colonies of Noctule Bats, Otters, Badgers and Water Shrews, plus a great selection of woodland birds. As if that was not enough, the site had a huge reed bed and a lake. Sitting in the hide, I started to watch a host of Wigeon, Gadwall and a few Mute Swans. Without warning a Bittern leapt out of the reeds and took a few flaps before flopping back down and out of sight. It was amazing to think that this species was resident here.

Like many urban sites, this special reserve is under severe threat and is in danger of losing its remaining surrounding green corridor, the buffer zone for the reserve's wildlife, to the lunacy of council planners. There is an active ongoing campaign by local residents and NGOs to try to halt the proposed development. It is indeed a nationally important urban reserve.

When I arrived at Washington WWT I was a little heavy hearted. Urban birding sites are all too often under the dark omnipresent shadow of development. That is why it is so important for us to recognise these areas and fight to get them protected. I sorely needed a pick-me-up and that is what I got when I entered a hide to witness the spectacle of several hundred Curlews flying in to roost. They looked and sounded amazing in the evening gloom. I have never seen so many Curlews inland before. Just as I thought that I had seen it all a Barn Owl flew right by the window to land by the side of the building. It had pounced on an unfortunate rodent and had its wings spread out across the grass. A brilliant end to a fantastic day.

Norwich, Norfolk

It's funny, but when I mentioned that I was going to Norwich to my non-birding friends in London it solicited the predicted chorus of country-bumpkin accents peppered with Delia Smith and tractor references. As expected, the birders I spoke to about the provincial capital of Norfolk suggested that my time would be better spent scanning for rarities on the north Norfolk coast. It was a tempting thought.

Until recently I saw Norwich as a gateway to that birding Mecca of the north Norfolk coast. In my early birding days I would go to bed at stupid-o'clock, wake up in the wee hours and drive up with my mate for our customary Norfolk birding weekends. We would set off in darkness and by the time we passed Norwich dawn would be breaking. We knew then that we had arrived in Norfolk. This time instead of passing by, I was going to stop and take a look.

While many Norfolk birders start their day in a marsh stuffed with birds I started mine on a council estate – where else? I was in Dussindale in the southern suburbs. My quarry? Not errant joy riding East Anglian hoodies but a flock of roving Waxwings. I cruised the early-morning, quiet residential streets like a binocular-clad curb crawler, peering through the windscreen into the front gardens of unsuspecting Dussindale citizens. Within minutes I was watching fifteen of these northern berry-gobbling beauties settled on the bare branches of a tree in a close. They trilled merrily as I sat in the car beaming, thankful that Waxwings considered themselves urban birds.

I then travelled to Mousehold Heath, a very historic area in the north-east of the city. Dubbed as 'the country in the city' by locals, its 200 or so acres are largely composed of woodland, though a small heathland area still exists. This site attracts a healthy breeding population of the more common warblers, though in days of old it also held breeding Wood Warblers, a species that has all but disappeared as a breeder in East Anglia. As it was the middle of winter, my visit resulted in very little other than the obligatory roving Long-tailed Tits, but I could see that this site had the potential to be fruitful at any other time of the year.

News of a Firecrest that had been lurking in bushes alongside the River Wensum near the city centre came through and temptation got the better of me. A short while later I was standing by the Wensum watching a Kingfisher quietly perched on the opposite shore near the Cow Tower, a ruined military tower dating back to the 14th century. I began to systematically search the bushes and trees that fringed the buildings by the riverside path. My attention was taken by a group of Black-headed Gulls that seemed to be playfully flying up and down the river. At one point around twelve of them landed on the water in front of me and among them was a single Common Gull – which I've seldom seen swimming.

Then I heard the characteristic sounds of an approaching tit flock and soon there were about fifteen Long-tailed Tits with smaller numbers of their more colourful Blue and Great cousins among them. As luck would have it, I also heard the distinctive 'peeping' call that could only have been a Firecrest. On cue, the little mite flew over my head and I just managed to catch it in my bins as it swooped up into a riverside tree. I heard it call once more as the foraging tit flock moved through, but despite waiting for it to reappear I didn't connect with it again.

One of Norwich's most famous birding spots has to be the UEA Broad in the grounds of the University of East Anglia. The Broad is, in reality, a lake with reedy fringes, and has accumulated a pretty impressive list over the years. It's a magnet for waterfowl and is a regular spot for wintering Goosander. Indeed, on the day I visited three glorious males and a female were on full view on an ice-free patch. I never tire of watching this handsome sawbill. I scanned through the milling gulls that stood nervously on the frozen ice by the shore. I was hoping to discover another Ring-billed Gull after the Broad's famous first record, but had to be content with the usual array of the commoner gulls. The UEA Broad certainly looked like a site that deserves regular coverage, despite its many dog walkers.

I ended my Norwich odyssey with a visit to Whitlingham Broad. Again the term 'Broad' was a bit of a misnomer as it was an old gravel pit. The site held Norfolk's second ever Black-and-white Warbler in November 1996. I stood by the trees it frequented and in the evening light tried to imagine this Nearctic mega foraging in its unique Treecreeper-like fashion. What a sight that must have been for the gathered hoards that came to twitch it. Suddenly, a party of noisy roving tits roused me out of my daydream and, homage paid, I turned to scan the Broad. Almost immediately I found a small group of Tufted Ducks and among them was a female Scaup – a superb sighting.

Would I have seen better birds had I gone to the north Norfolk coast? You know, I'm not so sure.

Peterborough, Cambridgeshire

It was a cold, misty grey Sunday morning in January. If the sun had risen it was keeping it quiet. In a small house on the edge of Peterborough, Cambridgeshire, was *Bird Watching* magazine's very own Mike Weedon (assistant editor), sitting alone at his breakfast table slurping coffee and munching on toast while waiting for a guest to arrive. This guest was due any minute and the plan was for him to escort this person around some of the finest birding establishments the town could muster. The doorbell rang. It was 7.35 a.m. He's five minutes late, Mike thought to himself as he strode to the front door. What he saw before him when he opened the door caused his eyebrows to rise impossibly high and his jaw to involuntarily drop open. There on his doorstep in the light drizzle was a tired, shoeless and pennyless birder. That birder was me.

An hour previously in the nearby village of Castle Bytham I had risen from my bed with a start having not heard the alarm and was frantically trying to find my trainers. My hosts had very thoughtfully 'tidied' them away somewhere and I didn't want to wake them or their small child at that ungodly hour. With the clock ticking and not wanting to be late for Mike I left the house, padding across wet grass and a damp road in my socks to get to the car. I was literally minutes away from his place when I realised that I had left my wallet back at the house. Could things get any worse?

Thankfully they didn't. In fact, it got a whole lot rosier by the time we had arrived at Maxey Pits, our first port of call. I had found a pair of wellies in the boot and all was well in the world. Maxey Pits, not to be confused with Maxi Priest, as I kept on wanting to call them, are a series of gravel pits that in true gravel-pit fashion are attractive to waterfowl and waders. The pit we visited was Mike's local patch, situated off Etton Road, and was the only pit in the complex that was managed for wildlife. It contained rather interesting looking reed beds, marshy areas and riparian woodland as well as open standing water. As we squelched around in the mud circumnavigating the ridge of the pit, Mike pointed out the areas that historically were reliable for the first returning Wheatears and Whinchats. The reed beds attracted breeding Sedge and Reed Warblers as well as Reed Buntings, plus the rough grassland area boasted a few pairs of nationally scarce Yellow Wagtails. The latter species is a bird that I have rarely seen anywhere during the breeding season in recent years.

It was the waders and the waterfowl that stole the show for me on our visit. There were quite large flocks of Teal, Shovelers and Tufted Ducks, with a good showing of Mallards, and along the shores were mostly Lapwings with a sprinkling of Redshanks and a couple of Dunlins. During the breeding season small numbers of Lapwings, Redshanks, Oystercatchers, and Ringed and Little Ringed Plovers all choose to raise their families here, and during passage periods you stand a fair chance of discovering something

interesting among the more expected shorebirds. Pectoral
Sandpiper, Buff-breasted Sandpiper and a Black-winged
Stilt a couple of years ago have all shown up. Mike proudly
told me that the site has also recorded Richard's Pipit,
Green-winged Teal and Glossy Ibis. It just goes to show
that if you cover a patch regularly enough, no matter what
it looks like and where it is, you will eventually uncover
gems. As we headed back to the car we took a quick detour
to scour a marshy reed bed in the vain hope of seeing a
reported Jack Snipe. Within minutes a solitary bird towered
out of the boggy ground it was standing on to relocate a
few yards away. Nice.

My time in Peterborough was short as I had to be back
for lunch, so we popped into Ferry Meadows Country Park.
A very picturesque area, it is known as the Ortons and lies
on the western edge of Peterborough near the offices of
Bird Watching magazine. There was a good blend of habitats,
ranging from woods to pastureland, which draws in a good
spread of species. It is also a very popular area with humans.
The River Nene flows through it and must look lovely on a
beautiful summer's day, but on this particular visit I had to
use a lot of imagination as we stood under grey, heavily
laden clouds. After crossing the attractive Milton Ferry
Bridge, we approached the three big flooded pits: Gunwade
Lake, which is used for sailing and watersports, Overton
Lake and the smaller Lynch Lake. All are great places to look
for the sometimes extraordinarily approachable wildfowl,
grebes and gulls. April and May are good times to come
here to watch out for Little Gulls, and Arctic and Black
Terns and, if you're even luckier, you might catch up with
the odd Sandwich or Little Tern. In the scrubby areas
surrounding the lakes roam breeding warblers, including
Lesser Whitethroats, Grasshopper Warblers and a few pairs of
Spotted Flycatchers – another species that I almost never see
at breeding sites. It is clearly a good place to cover and as
Mike emphasised, anything can turn up at anytime, as

illustrated by the Red-rumped Swallow that materialised during spring 2010.

What a great introduction to a city that many of us would never equate with birds. After bidding Mike farewell I journeyed back to Castle Bytham to search for another elusive rarity: my trainers.

Plymouth, Devon

I was thinking the other day that visiting the cities of Britain and beyond must be similar to being a judge on *The X Factor*. Each city is almost like a contestant being judged on its ability to attract urban birds and urban birders. But unlike in the TV series, in this competition everybody is a winner. Indeed, the objective is to find as many positive points as possible, and no city has come anywhere near failing. So far I have visited dozens of cities with an additional few not-so-urban idylls, and I'm yet to find a city that offers no birding talent. But how can you judge a city when it tips down with rain the whole day and the birds have sensibly taken shelter. I faced that awful dilemma on my latest urban birding foray in Plymouth recently. After experiencing the driest spring since the devil was a boy, the heavens decided to open and pour every liquid ounce of rain that they could muster onto my head.

I had planned a nice early start with my host, West Country expert and twitcher extraordinaire Sara McMahon. Our plan was to first check her garden moth trap pre-dawn before hitting Plymouth's premier birding sites. The previous night I had bullishly insisted that despite the forecast we would wake up to a dry morning and sunny afternoon. I was made to eat those words as we drove around the deserted streets of Plymouth in the chucking rain. We meandered from the east of the city to the west, beginning our wet adventure in the car observing Hooe Lake through swishing windscreen wipers. The 'lake' is in an inlet off Plymouth Sound and in fact is split into two lakes: one tidal and the other fresh water. Although we didn't see much in the way of avian activity I was reliably informed that the site was generally good for Little Gulls at almost any time of year, plus wintering duck such as Shelduck, and grebes and waders. Nearby Mount Batten seemed like a nice place to take a walk, had it been a sunny day. There was also an inviting walled promenade jutting into the sound that is a magnet for the local birders. Depending on the wind direction, one side has calmer waters while the other holds back rougher seas. Sara informed me that terns, all three divers, Gannets and Fulmars are certainties during the appropriate seasons. Meanwhile down the road at Hooe, previous winters regularly brought in Purple Sandpipers, Black Redstarts and, exceptionally, Bonaparte's Gull and even Ross's Gull twice.

After about four hours we decided to retreat back to Sara's to dry off and indulge in some tea and biscuits. I decided to take the unprecedented step and go for a siesta. What else was I to do on a wet Sunday afternoon? When I surfaced again some two hours later the rain had eased off significantly. However, as soon as we walked out of the house the rain started. Undeterred, Sara and I set off, briefly visiting the National Trust-run Plymbridge Woods which resembled a tropical rainforest complete with a raging river running through it. Aside from the usual woodland suspects, it also has breeding Nightjars. When we arrived at Ford Park Cemetery, a break in the clouds allowed us to actually leave the car. Very close to the city centre, it's thirty-five acres of gloriously wooded open space carefully landscaped and clearly of benefit to the local wildlife. Blackcaps and Chaffinches sang, Swifts swirled overhead, and a couple of Ravens took to cavorting around some headstones with attendant Carrion Crows and Magpies. Sara had discovered a singing Common Redstart a couple of days previously that could not be found during our visit. For me, Ford Park Cemetery looked like a great migrant trap with many sheltered bushy areas, and indeed historic records of Wryneck validated my thoughts. In the winter, Black Redstarts, Bramblings, Siskins and the winter thrushes should be looked for, while Hawfinches have also been turning up over the past couple of years.

The drizzle recommenced as we left the cemetery and by the time we were heading into the centre of town, motoring alongside the River Plym, it started bucketing again. Distantly, from the corner of my eye, I noticed the unmistakable form of a tern hawking over the river. It was clearly smaller and daintier than the Black-headed Gulls it was loosely associating with. I alerted Sara and stuck my neck out by calling my mystery tern a Little Tern. She expressed surprise, as any species of tern in June on the Plym was a definite rarity, and quickly drove us back to the spot where I had last seen it. After some twenty minutes of scoping the river in the driving rain we eventually relocated my bird. It was a Little Tern. We skipped with joy like two drenched puppies. This dainty species is a particular rarity during the summer as the nearest breeding colony is probably more than 100 miles away in Dorset.

Unbelievably, despite it being a complete washout, I had a great day in Plymouth. It was plain to see that the city is a great venue for urban birding with an impressive species list. It certainly has the X factor. Needless to say, the following day when I was back in London, Sara called me and proclaimed: 'Where are you? It's a beautiful sunny day here in Plymouth.'

Sheffield, Yorkshire

We've all wandered past a housing estate or shopping centre and remarked: 'I remember when that was all fields'. How many of us can gaze at a landscape and say: 'I remember when this was an open-cast mining pit'? Well, that was exactly the case when I accompanied Sheffield birder Mark Reeder for a tour around his city on a freezing cold day in January. The whole area had suffered terribly from the deluge of snow that hit the UK during the winter, causing my planned visit to be postponed several times. In places the snow was thigh deep, and although it had melted by the time I made my appearance there were still sullied piles of the stuff on some street corners. Out birding, the walking conditions were sometimes treacherous, with thick ice and, worse still, black ice still prevalent.

But the sun was shining and the sky was blue when we strolled into Pit House West at the northern edge of the Rother Valley Country Park. To my eyes it looked like a natural enough mix of scrubland and riparian habitats but it was in fact a redeveloped mine. Had we visited during the summer our ears would have been ringing to the jittery notes of Reed Warblers as this is certainly the best site in Sheffield for this common 'Acro'. The occasional Grasshopper Warbler stops to raise its young, but it is probably the Willow Tits that most local birders come here to find. Our understanding of the plumage differences between Marsh and Willow Tits has recently been thrown into confusion, with some experts now saying that the only sure-fire way of separating the two is by call. Well, you wouldn't have to worry here, as Marsh Tits are unknown. Mark showed me the frozen reed-fringed pool where their wintering Bittern had been feeding out in the open, but I had no such lucky encounter. One or two of these secretive reedy denizens have been wintering at the site for the past eight years.

There are several locations within the city boundaries that have been beautified and made good for wildlife after previously being putrid pits devoid of life. Astoundingly, despite the obvious work and effort that has gone into naturalising these areas, the threat of development still looms. Pit House West is earmarked to be Sheffield's next leisure centre, while over at Orgreave Lakes, part of the land has already been marked off as a new housing estate. The lakes themselves are a new addition to the birding map, having only been in existence for a couple of years. To me it looked like a landfill site comprising a main lake and a smaller satellite lake surrounded by freshly bulldozed mud next to the River Rother. But it was exhilarating for Mark, as it was his local patch where he had the rare opportunity to record its birdlife almost from day one. As we crossed the cascading river to get into the site, a few Goosanders warily swam away from us, riding the lapping waves almost in the way I would imagine a Torrent Duck would beat a hasty retreat on a Chilean river. Parties of skittish Teal, Tufted Duck and a couple of Goldeneyes took flight, leaving a fair number of Mallards behind. The normally docile Mallards were torn as they nervously looked at each other wondering, 'Shouldn't we be taking off too?'

The lakes were largely frozen, forcing nearly ninety Wigeon and a similar number of Gadwall out onto the muddy shores and surrounding fields – a vision that left Mark swelling with pride. As we sifted through the ensemble of loafing gulls gathered on the frozen lake looking for a recently sighted Caspian Gull, like any proud patchwatcher, he recounted some of the star birds he had found at this developing gem – for example the recent Snow and Lapland Buntings brought in by the inclement weather systems. Most surprising was the Leach's Petrel he discovered flapping around the lake last September. As he watched the bird begin to gain height a Sparrowhawk zoomed in out of nowhere. Thwack! Mark's site tick was no more.

No city visit is complete without checking out the real centre of town. On the way there we drove past Bowden Housteads Woods, Sheffield's oldest beechwoods that have been in existence since the 17th century. Although Lesser Spotted Woodpeckers have been reported here and a Yellow-browed Warbler was found a few years ago, the woods are currently woefully underwatched. The same may not be said for Sheffield Botanical Gardens, our inner city destination. At nineteen acres it was a fairly small park but it was rammed with birds. Tit flocks proliferated with a supporting cast of Chaffinches, Nuthatches, a Treecreeper and a Great Spotted Woodpecker. The gardens' claim to fame was the 1987 Black-throated Thrush, plus more recently an Arctic Redpoll and a Yellow-browed Warbler. The city centre also boasted the obligatory pair of Peregrines that are usually to be found surveying their kingdom from a lofty perch on St George's Church or the nearby BT Tower.

Sheffield is a great place for urban birding, with Dippers and Kingfishers breeding in the city centre and a population of suburban Black Grouse, albeit introduced.

Southend, Essex

It's funny that sometimes we can live relatively close to an interesting birding location but never consider visiting it. Instead, we find ourselves travelling for miles to distant sites drawn by the perceived promise of better birding. Southend is not far from where I live: a mere forty miles from central London. If I left early enough on a weekend morning I could be there, sitting in my deckchair lapping up the Essex sun in under an hour. I was convinced that I had visited this east-coast seaside town before. Somewhere in the deepest recesses of my mind I had fond childhood memories of roller coasters, candyfloss and monstrous seagulls. However, after rigorously cross-examining my mum the truth emerged. I had never been to Southend before – the erstwhile number one seaside retreat for Londoners.

I recently made the journey there one autumn after responding to an invitation from local birder and born-and-bred Southender Emily Broad. She was keen for me to see some of the sites she birded and to help find her first-ever flycatcher, of any species. The gauntlet was laid down. Southend-on-Sea, to give the area its full title, is more than just a seaside town. The whole region is a slither of land with over ten miles of coastline that encompasses Leigh-on-Sea in the west and Shoeburyness to the east. To the average day-tripper, the view across the Thames Estuary at Westcliff-on-Sea, near Emily's home, would be somewhat blighted by the industrial landscapes of both Canvey Island and the Isle of Grain in neighbouring Kent. Yes, the view is an acquired taste but both spots are, of course, exceptional for birds and are places that I used to visit regularly as a younger birder.

Strolling along the Southend seafront the first thing we saw when we scanned the estuary were the multitude of gulls and waders variously resting and feeding on the exposed tidal mud. Little Egrets, Oystercatchers, a few Curlews and Turnstones stalked the foreshore augmenting the masses of large gulls, predominately Herring and Great Black-backs, which peppered the sludgy mud. Further out were eighteen Common Seals that were loafing on a distant mudbank. Heading towards the world-famous Southend Pier we picked up a few Common Terns investigating the saline channels, many Black-headed Gulls and the odd gorgeous Mediterranean Gull. The shoreline that we were watching was part of a protected Foreshore Nature Reserve – the largest of its type in England. The area has a rich mussel bed and is stacked with invertebrates, brilliant for its many winter-visiting waders and wildfowl, including big flocks of Brent Geese.

Southend Pier was a treat. It is the longest in the world stretching out over 1.34 miles into the Thames Estuary. Although it is served by a cute little railway, Emily and I decided to do the walk to the end in search of the Turnstones and Common Terns that often perch on the structure. Sure enough, we eventually found assorted flocks of both species resting up along with parties of terns fishing just out to sea. Some of the terns were still in full summer plumage but most were either immature birds moulting into their winter 'garms' or adult birds already in winter attire. Our attention was quickly drawn to two seemingly smaller winter-plumaged terns sitting in between two more classic-looking Common Terns. These odd birds superficially resembled the more familiar terns but they had spindlier, shorter legs and were definitely physically smaller. Emily and I took many pictures that we later posted on the Internet for comment. It resulted in much debate among my Facebook buddies with the general consensus being weighted towards Common Terns, albeit aberrant ones.

The end of the pier itself is a good place to be hunkered down during an autumnal easterly blow, as the seawatching can be excellent. Gannets, Arctic Terns and Bonxies are regular, and it is worth scanning the water's surface for Red-throated Divers, scarcer grebes like Slavonians and potentially interesting gulls. After our saunter to the end of the pier and some obligatory candyfloss we caught the train back to where we had started. Opposite the seafront were some low cliffs that were either fenced off scrubland or manicured sloping wooded parkland. We investigated some of the parkland, paying particular attention to bushes that afforded the most cover for migrants. My goal was to find Emily her flycatcher. I failed miserably and could only produce around ten Robins.

I found to my surprise that this oft-disregarded part of eastern England's coast is surprisingly good for birding. According to the website of the Southend Ornithological Group (SOG), there are at least forty-one birding sites in the area. Their vote for the best site went to Pitsea Hall Country Park. With a blend of habitats that includes a freshwater fleet and saltwater creek this site has pulled in a bag full of special birds. Aside from waders and a good showing of waterfowl, Water Rails and Bearded Tits are possible. The rarity list was not to be sniffed at either, with gems like Olive-backed Pipit and Great Reed Warbler.

We ended our day on another of the SOG's featured sites: Two Tree Island in Leigh-on-Sea. We visited the west side of this Essex Wildlife Trust reserve where Emily proudly showed me the Little Egret roost there. The reserve is a fantastic little spot with great birds and in my opinion well worth adopting as a regular haunt.

A week later, back in London, I received a tweet from Emily. She had found her own flycatcher – a Pied Flycatcher on Southend Pier. This further proves that we don't have to travel far to see good birds.

St Albans, Hertfordshire

When I explore cities for my magazine column I am usually accompanied by a local birder who is invariably either older then me or at the very least a similar age. A recent visit to St Albans blew that statistic out of the water. Luke Massey, my guide for the day, was a fresh-faced nineteen-year-old university student who when not studying and birding in Canterbury is to be found patrolling the birding spots that this ancient town has to offer. During the course of our day I found myself imparting pearls of wisdom to my young companion and not just about birding as he we was certainly a competent birder, but about life and girls. He sure did dig up some old memories.

When we met he was a little bit worse for wear after a big night, so we took it easy strolling around our first location, Verulamium Park. As you may have guessed from its name there is a Roman connection; indeed, the park is on the site of the Roman city of Verulamium, which was apparently established as a city around the same time as a certain Londinium. The park itself covers about 100 acres and includes some still-visible Roman wall relics. The River Ver that flanks the site to the east and the Verulamium Lake itself hold the usual common duck species spiced up with the occasional Teal, and the small woods on the island in the middle of the lake are home to a reasonable-sized heronry. The park has attracted Green Sandpipers – Common Sandpipers are a local rarity as was a recent Black Tern and Egyptian Goose. Luke explained that it is well worth scanning the skies above the woods on the north-east horizon, as you may be lucky to spot soaring Red Kites and Common Buzzards. We popped around to the headquarters of the Herts and Middlesex Wildlife Trust, which skirts the north-western edge of the park, to check out their small wildlife garden. Strangely, they are not open during the weekend, but we peered over the wall and saw Chaffinches and various tit species gorging themselves on the ample feeders.

I love visiting those tucked away little sites in cities that only a few locals visit and St Albans had a few of those. One such site was Riverside Road Watercress Beds, a short drive from Verulamium Park and literally a few hundred yards away from the town centre. We were greeted by a flock of Jackdaws loafing on the treetops and adjacent rooftops, and cooing Woodpigeons that immediately made me think of the summer even though it was a freezing cold day in January.

I really felt a sense of tranquillity when we walked into this tiny local nature reserve run by the Watercress Wildlife Association. It was a delightful area of damp woodland and small ponds, with the ever-present River Ver flowing through it. The Alders were alive with Siskins variously singing, feeding and drinking from small pools. We also found a solitary Lesser Redpoll in their midst, and no doubt there were more around. Water Rails were a distinct possibility according to Luke, although my heart sank slightly when I heard the familiar squawking of my old friends (said in the most sarcastic tone possible) the Rose-ringed Parakeets, a bird yet to gain a toehold in the area. In the summer the site boasts small numbers of the usual common woodland warblers and is a good place to look for Kingfishers.

No mosey around the St Albans area would be complete without a visit to perhaps its most famous site, Tyttenhanger Gravel Pits. It is well watched by birders and regularly attracts upwards of 140 species annually. Whenever I think of Tyttenhanger, I think of Ospreys, a raptor that does seem to appear annually there. It was my first time at the site so Luke introduced me to where the small Tree Sparrow colony was near Garden Wood on the eastern edge of the pits. It is apparently one of the most reliable sites for this nationally scarce species in the county. Of course, I didn't see any, but there were plenty of Yellowhammers and Reed Buntings converging on the feeders. The fields near Willows Farm, to the south, were alive with mixed parties of Redwings, Fieldfares, Lapwings, and Common and Black-headed Gulls all stalking the ground for tasty morsels. The thin tree cover leading from Willows Farm into Garden Wood is also well worth sifting through for the elusive Lesser Spotted Woodpecker.

The Herts and Middlesex Wildlife Trust's Lemsford Springs Nature Reserve was by far my favourite place in the town. Bordered by the River Lea to the west, Lemsford Springs is an old watercress bed surrounded by woodland, meadow and marsh. The lagoons were brimming with millions of freshwater shrimps that in turn attract a myriad waterbirds. We watched an obliging Kingfisher that posed beautifully on the end of a pole just outside the hide and beneath it a Little Egret that was stirring up food items with its foot. We even enjoyed a surprise Water Rail that swam in full view from one clump of reeds to another. But it's the Green Sandpipers that the site is most famous for. A well-studied bird here, many have been ringed and Trust volunteers have worked out that Green Sandpipers guzzle up to 8,000 shrimps a day.

Staines, Surrey

Sacha Baron Cohen has a lot to answer for. As Borat he famously offended the Kazakhstan Government to the highest degree, but less well known internationally is the angst being felt by all of 'da Staines Massive'. In an attempt to repair the perceived damaged reputation the town has suffered at the hand of Ali G officials renamed it Staines-upon-Thames. Will the proud people of Staines ever recover from this embarrassing episode?

When I recently returned to the Staines area on a freezing overcast December day the memories began to flood back – memories of being an eighteen year old in Wembley excitedly calling for my mate Alan, who lived around the corner, and driving to visit Staines Reservoir in his new old banger. We saw some great birds as a result of well-timed visits, including our first flocks of spring Black Terns, an eclipse Garganey and a first-winter Glaucous Gull. That was over twenty years ago, so my visit, escorted by veteran Colne

Valley birder Pete Naylor, was long overdue. In truth, I had been to the reservoir in spring 2009 to twitch a stunning adult White-winged Black Tern, but that visit didn't really count because I stood on the extreme western end of the causeway for a mere thirty minutes before having to rush back into London for a meeting.

Staines Reservoir is perhaps the most famous birding spot in the district, with a bird list longer than most people's arms. Built in 1901 it is the oldest reservoir in the region and over the ages has been the stamping ground for many people of the London birding scene. It is, in fact, two reservoirs (north and south basins) that lie south-west of Heathrow Airport and are usually under the shadow of a noisily landing jumbo jet. A central causeway that has general public access separates the water bodies, and the site is a Site of Special Scientific Interest (SSSI) due to the important numbers of wintering diving ducks. In the days before bird-information services and when the grapevine was king, the causeway was a great place to meet up with fellow birders to chew the fat and to find out what rarities were around. In the winter you were exposed to arctic winds with no cover and in summer your own personal cloud of midges would envelop you. I used to wish that I had an attendant gang of Spotted Flycatchers to take care of the pesky flies. Regardless, a remarkable array of rarities has occurred here, including a host of waders attracted in when one of the basins is drained for maintenance. Beauties such as Buff-breasted Sandpiper, Collared Pratincole and Wilson's Phalarope have all turned up, while interesting terns and gulls are regular. It can be quite an exciting place on the right day as you can often see migration in action as you watch waders sweeping in.

The Staines area falls within the jurisdiction of Spelthorne Borough Council, and if you were to take an aerial view of the parish you would see a patchwork of reservoirs, flooded fields and other riparian habitats, including the River

Thames. Not all the areas are accessible to birders due to the close proximity to Heathrow Airport. As you can imagine security is extremely tight in places post 9/11, with many of the places that I used to sneak into, such as King George VI Reservoir adjacent to Staines Reservoir, strictly out of bounds to the casual birder. A special permit has to be obtained from Thames Water. My guide Pete had a permit so we took a quick stroll up to the bank for a sneak preview and were treated to parties of waterfowl that included Tufted Duck, Goldeneye and a solitary female Velvet Scoter. It was the first time that I had seen this normally maritime duck in London.

Next door to the west, bordered by the busy M25 and split by the River Colne, lies SSSI-designated Staines Moor, recently famous as the temporary home of a much-admired wayward Brown Shrike. At 1,275 acres, it is one of England's largest areas of neutral grassland, and it contains the oldest-known anthills of the Yellow Meadow Ant in Britain. The terrain is littered with these strange structures, some of which are over 200 years old. Pete informed me that the usual expected warblers breed there, while in winter the area and the adjoining Stanwell Moor attract winter thrushes, Snipe, Jack Snipe and the occasional Barn Owl.

There is a host of other nearby sites that are far less frequently birded than the sites that I have already mentioned. The majority are gravel pits, including Hithermoor Lake in the north-east corner of Stanwell Moor, which is apparently fairly good for wintering duck and Water Rail. Just outside the borough on the western side of the M25 is Wraysbury Gravel Pits. In my day it was a great winter spot for finding Smew. Pete confirmed that this was still the case, though the classic numbers reported over twenty years ago were sadly a thing of the past.

Have I convinced you that Staines is a top birding spot yet? My message to the people of Staines is that there was no need to change the town's name. Staines' ornithological legacy is legendary and that will never be tainted.

Stockton and Middlesbrough

One evening I watched a distant Short-eared Owl hunting under a dimly lit and depressingly murky grey autumnal sky. Only this was not autumn, it was during the middle of our glorious English summer: an evening in June to be exact. It was hardly balmy as I felt a tad cold watching the owl through my scope flapping leisurely over the edge of a field and up an embankment. I tracked the bird as it suddenly pounced on something that must have been unfortunate enough to have caught the owl's eye or maybe its ear. As we know, owls have superimpressive hearing to complement their exceptional sight. It disappeared behind a bush. Satisfied with the fascinating spectacle that I had witnessed I began to scan the horizon on the lookout for any other potentially interesting birds.

You might now be imagining me scanning some expansive moorland or beautiful marshy estuarine vista. You would be wrong. I was standing alongside the River Tees in Teesmouth National Park surrounded by a highly industrialised hinterland north of Middlesbrough and very close to Saltholme RSPB Reserve. The landscape ahead of me was punctuated by a multitude of high-rise pipes and flame-throwing funnels. Ever seen the opening sequence in Ridley Scott's acclaimed *Blade Runner*? This very landscape directly influenced the weird architecture Deckard, Harrison Ford's character, traverses through. Who said ugly can't be beautiful?

I ended up in this strange place at the culmination of a long day exploring the wilder side of Middlesbrough and Stockton in the company of Steve Ashton, Tees Valley Wildlife Trust's People and Wildlife Manager. Situated in the principality of Teesside, this industrialised urban centre is largely ignored by everyone ranging from politicians to non-resident birders – unless there is a rarity in town. Middlesbrough and Stockton are almost like twin towns separated by the River Tees, and the urban birding was surprisingly varied. Take, for example, Portrack Marsh in Middlesbrough. It was a very interesting looking fifty-acre remnant marshland reserve managed by the Tees Valley Wildlife Trust and Northumbrian Water. The ever-present River Tees borders the site and during the summer Sand Martins and Kingfishers nest along its raised banks, and at the right time of year it hosts migrating Salmon. Despite being encroached upon by industry, the reserve's mosaic of open water, reed beds and scrubland pulls in some good birds. The usual riparian warblers breed alongside Reed Buntings, while Water Voles are resident. During the winter it receives the expected array of waterfowl, and small gatherings of Grey Herons. Scarcer ducks are sometimes to be found, including Smew and Long-tailed Duck. However, the local twitching fraternity got very excited when two male Penduline Tits showed up in 2006. This nationally scarce annual visitor is exceptionally rare this far north.

Bowesfield Nature Reserve, however, was a very different proposition. Originally farmland, it was left to go wild in 1996. By 2004 development plans were afoot to use the land for housing and commerce. The developers, to their eternal credit, decided to work with the Wildlife Trust to turn part of the land into a nature reserve, with a series of small pools and open lakes leading to the river interspersed with wildflower meadows. The new residents were also offered free membership of the Wildlife Trust. It was a superb opportunity to engage people who may have otherwise viewed the reserve as waste ground. Indeed, the very same good people would have been first on the scene to witness a recent spring Red-rumped Swallow swooping over the heads of the more regular Skylarks and Reed Buntings. I love it when a plan comes together.

Steve and I popped into a few other interesting, though thoroughly urban, spots like Middlebeck in the heart of a Middlesbrough housing estate. Its attractions include Water Voles, Reed Warblers and even Grasshopper Warblers. The luxuriantly wooded Linthorpe Cemetery was also quite a surprise in the middle of the urban sprawl. It harboured the classic woodland birds that you would expect to find, though it is very underwatched.

I ended my urban exploration twitching a Nightingale that was singing way north of its normal southern range at Cowpen Bewley Country Park in Stockton. This former landfill site is now part of the Tees Forest Development. Needless to say, when we arrived at the spot it had decided to clam up and not utter a single note.

Undoubtedly, Stockton and Middlesbrough's pièce de résistance has to be Saltholme RSPB Reserve. Despite having a backdrop of oil refineries and factory towers, it was certainly a marshy wildlife oasis. I was shown the lie of the land by ultra-enthusiastic assistant warden Toby Collett. During the time I was there I managed to dip out on a Roseate Tern but caught up with a Mediterranean Gull, a first-summer Arctic Tern that Toby expertly picked out, a female Marsh Harrier and a Barn Owl, which we watched wafting into an empty building. The reserve has amazing birding credentials: from county notable records like a booming Bittern, Purple Heron and Glossy Ibis to national rarities like White-rumped Sandpiper and Pallid Harrier.

I was totally taken by my day birding in Teesside. Not only were there some very interesting places to explore but

I also loved the fact that here, too, people engagement was viewed as an imperative element in the region's conservation efforts. If you are ever in the area, come and get involved in some real urban birding.

The Olympic Park, London

It is well known that I am a big advocate of watching birds and nature in cities, constantly waving flags and banging drums about how exciting and rewarding urban birding can be. Some non-birders that I speak to don't get the concept of urban birding let alone the existence of urban wildlife. To them, the mere mention of cities and wildlife in the same sentence would be like mixing water with oil. But as we all know, in reality that is far from the truth.

Herein lies the dichotomy. Many of the urban birding locations that I visit are essentially relics from the past; islands of natural habitat like woodland, rivers or grassland that have somehow survived despite being touched by the hand of human interference. Some of the places have no official protection and are viewed as oases by the resident birders and waste ground by developers. If your cherished local birding paradise gets churned up by monstrous JCBs to become a shopping centre it is hard not to feel anger, resentment and sadness. But is it possible to seemingly destroy an urban area that was once productive for wildlife and replace it with structures of steel and concrete and still retain its attraction to nature?

It's hard to imagine just how such a proposition could work, even for an eternal optimist like myself. So when I was invited on a site visit to the Olympic Village, in my mind's eye I had the vision of a huge building site crawling with shouty people in hard hats. The Olympic Village lies at the southern end of the Lea Valley adjoining Hackney Marshes in east London. It was a contentious construction project that faced criticism from many different quarters, not least from a tiny number of birders who used to cover the site before the area's makeover. Previously, it was part of a large expanse of derelict land littered with decrepit factories, greasy garages and unkempt scrubland. The birders I contacted nostalgically spoke about regular sightings of Cuckoo, Snipe, warblers and Song Thrushes. Even I remember the area as an eighteen year old visiting a girlfriend in nearby Stratford. In those days, while on the way to her council estate avoiding dodgy feral dogs – and people for that matter – I would spend a brief moment scanning the expanse of waste ground. I too was once blessed with the view of a passing Cuckoo.

Inviting as this past scenario might have been, the cold truth was that most of the vegetation consisted of invasive species like hogweed, and hidden under the soil were tons of contaminated toxic waste. This potentially dangerous cocktail had to be removed, according to David Stubbs, head of sustainability at the London Organising Committee of the Olympic and Paralympic Games. A keen birder, he took me for a stroll around the site along with two of his colleagues to introduce me to the Olympic Park. From day one, the vision was to build a complex that fused sustainable biodiversity with human needs. To be fair, I was very impressed with the level of care and thought that had gone into the landscaping and naturalising of the wild-looking and imaginatively named Olympic Park, which is contiguous with the Olympic Village. It seemed as if they had consulted with practically every environmental body under the sun during the planning stages. Even the actual village itself, with its low-rise buildings and stadia, was designed with nature in mind.

More than 2,000 hand-picked native trees were planted for both the wildlife and the hordes of visitors to enjoy. Many of the structures, including the bridges that span the river, were fitted with nesting holes for species as varied as Kingfishers, Sand Martins and bats. They even have a manufactured Otter holt in place. And as if to prove the point, several Sand Martins were indeed busying themselves swooping over the river on my visit.

The Wetland Bowl, the riverside woodland and reed bed in the northern part of the park, was established using vegetation sourced from East Anglian fens. Although small, it looked very promising. I was told that it had already attracted Kingfishers and singing Reed Warblers, and I noticed a Little Grebe discreetly slipping into a channel in the reeds. I joked that it might attract Bearded Tits and Bitterns, but on reflection I wouldn't be surprised if it does in reality. Despite all the construction work, there was quite a lot of avian activity overhead. Every time I looked up there were gulls aloft plus a regular Cormorant or two passing through. As we walked in the shadow of the now-iconic Olympic Stadium, Lesser Black-backed, Herring and Black-headed Gulls swirled around its circular roof. A Stock Dove flew by, cutting across the classic central London backdrop of Tower 42 and the Gherkin. I was getting the urban birding buzz. Bizarrely, a Short-eared Owl spent a short while flying around inside the stadium one winter. Perhaps its occurrence was a legacy of the past when this species might have ranged undiscovered across the general area.

During this visit the park was still under construction and in need of bedding in. By the time the Olympics were over, at the end of the summer in 2012, it had matured and was accessible to the public. I left the Olympic Park with one thought burning in my mind: after the world's athletes have finished competing, would there be Olympic birders limbering up to take on the challenge of watching this unique patch?

Tower 42 and Canary Wharf, London

I am delighted that the art of visible migration watching is now so in vogue. I absolutely love the idea of watching birds in the act of migration, especially from high vantage points, and have done for some time now. My first recollection of 'vis migging' was as a six-year-old total birding novice inadvertently watching the movements of birds in the early-morning urban skies. In those days I thought that the Lapwings I observed flying over during July, sometimes in good-sized flocks – now sadly a thing of the past – were simply randomly moving around. A couple of years later I discovered that birds actually migrated, but I thought that this occurred in one big movement almost exclusively along the coasts during April and September. I was soon to be proved wrong.

I began to read books by Eric Simms, one of my birding heroes, discussing falls of migrants in urban areas that were not too far away from me. He also spoke of early-autumn visible migration watching back in the Fifties and Sixties, when old-school birders used to stand on Primrose Hill close to Regent's Park to watch the procession of birds coming in from the north and moving across the urban vista. Those pioneers discovered that there was an obvious migration flyway through the 'North London Heights' traversed by hoards of Meadow Pipits, Chaffinches and other passerines.

One recent October Saturday morning, during my stint in goal at football, I was at the receiving end of a bad challenge by the opposing striker. After the game we got talking and I discovered that he was a cameraman and had just been filming on top of Tower 42 in London's Square Mile. He could not stop raving about the view and the gorgeous sunrise. I got in contact with the Tower the following week to ask if I could bring up a small crew to film Woodpigeon movements. To my surprise they agreed, though on the day we could only summon around 200 Woodies through the grey murk. I immediately saw the potential of this remarkable vantage point and a subsequent meeting between the Tower 42 management team and myself resulted, to their eternal credit, in the formation of the Tower 42 Bird Study Group.

The journey onto the roof of Tower 42 is not as straightforward as you may imagine, as there is no lift to comfortably take you all the way to the top. Instead, you have to make the final ascent on foot, climbing the fire-escape staircase, then up an even windier one into the plant room. In that area there are the pipes, platforms and low ceilings to negotiate, and finally a climb up two sets of vertical ladders before you can haul yourself onto the roof. Clearly, this is a journey not suitable for those of a delicate disposition.

The Bird Study Group met once a week during the spring for a total of nine sessions and recorded a fascinating collection of species that included daily Peregrines and Sparrowhawks, as well as central London scarcities like Oystercatchers, Arctic Terns, Common Buzzards, Red Kites, Hobbies and, most famously, a couple of Honey-buzzards. One of the HB's even managed to crash into a west-end office window, much to the alarm of the office workers within. Fortunately, the bird was unharmed and eventually resumed its journey.

Although extremely hard work at times – akin to seawatching but with the waves being replaced by rows of buildings disappearing into the horizon – Tower 42 has quickly become synonymous with visible migration in London. The autumn sessions have resulted in several Sandwich Terns and another Red Kite.

A few miles east lies Canary Wharf with its iconic One Canada Square, one of Britain's tallest buildings complete with its characteristic pyramid. Between 2001 and 2006 urban birder Ken Murray and his colleagues did a study of the migrants that occurred in the tiny Canada Square Park at the foot of the tower and nearby Jubilee Park. The latter park is an area of landscaped gardens filled with introduced evergreens, while Canada Square is a Spartan, thinly tree-lined lawn. Both areas are used by the multitudes of office workers as a place to consume lunch and fags. Despite their initial unattractiveness, Ken and his friends discovered an inordinate number of migrants in the parks over the years, including rarities like Red-backed Shrike, Wryneck and possibly as many as three Blyth's Reed Warblers.

I visited the area several times and immediately saw the potential. This has led to the re-formation of Ken's study under the guise of the Canary Wharf Migrant Bird Survey. Open to all, provided a permit is obtained (as security is tight), this potentially fascinating project could shed even more light on bird migration in built-up areas. At the time of writing the group had already recorded a Wheatear and a couple of Firecrests. Urban birding has truly come of age and has never been so exciting.

Wormwood Scrubs and Beddington Farmlands, London

While travelling the length and breadth of Britain and mainland Europe in search of urban birds, my trusty rucksack has adorned the floors of many hotels and has been slung about in transit on planes, trains and even a few cars. But going away at least once a month is not good when you are also trying to cover your local patch, especially during the migration period when your undivided attention is crucial. So I decided to stay at 'home' for an autumnal check on Wormwood Scrubs, my local patch in west London, and Beddington Farm, my former stamping ground during the Eighties in south-west London.

My staycation started with a visit to my surrogate ancestral home, Wormwood Scrubs. Literally ten minutes from my house, this park has really weaved itself into the core of my being. Every time I visit it I am filled with an inexplicable excitement and sense of expectation as there is often something unusual to be seen, however small. It's my own inner city Fair Isle – albeit with an infinitely minuscule fraction of the rarities. Even now, some eighteen years after I first discovered the place as a birding venue, I still receive strange looks from people when I suggest the prospect of birding there. It still suffers from the stigma that the prison attracts, and funnily enough the other day a birder asked me in all seriousness if I was a prison warden.

The Scrubs encompasses 183 acres of playing fields and grassland encircled by a thin band of woodland dominated by Sycamore, birch and plane. Quite frankly, at first glance it may not look like the sort of place that would attract birds. To the north are busy railway lines, to the east and west, housing and to the south the grounds of the infamous imposing prison. But despite this, the area has an incredible bird list with more than 120 species recorded since 1980, including Wryneck, Ortolan and two separate Richard's Pipits. It also houses a colony of breeding Meadow Pipits, the closest to central London. Up to five pairs of this amber-listed pipit eke out a precarious living in the grassland that is all too frequently breached by dog walkers.

It's mid-August and I'm yet to register a passage Whinchat, a bird that has come to symbolise all that is autumn in this part of the world. You can be guaranteed the presence of these fine-looking chats especially during the autumn migration periods. When I first started birding here numbers were exceptional, with one late-September day resulting in a count of twenty-two birds – the biggest group reported in southern England that autumn.

I had a good time checking bushes and managing to connect with quite a few Whitethroats, Blackcaps and several Willow Warblers. Then in the grassland on a thistlehead among at least 200 juvenile Goldfinches I found a solitary Whinchat. My heart swelled.

The next day I was in Hackbridge, deep in south London, at the invitation of Beddington Farm's patch birders Peter Alfrey and Roger Browne. The farm is a working site with gravel extraction, landfill and sludge spreading. It also contains flooded marshy pits that are good for passage waders, and the landfill site can attract hoards of gulls that can contain 'white-winged' scarcities including the mega Glaucous-winged Gull that seemingly commuted from Gloucester in 2007.

Walking around, I couldn't believe the transformation the site had gone through, though it looked very much like work in progress as the area was clearly in need of some good habitat management. At nearly four times the size of the London Wetland Centre, Beddington Farm really is a sleeping giant in London's ornithological world. With over 250 species on its list, of which more than 150 are seen annually, few sites within London's boundary can come close. Amazing birds like Killdeer, Spotted Crake (which also bred here in the Sixties), inland Sabine's Gulls, Tawny Pipit, and Rustic and Little Buntings were among the avian glitterati to have visited the area over the years. It is also home to one of Britain's largest breeding populations of Tree Sparrows, with some 300 youngsters fledged in 2008.

These were some of the facts that I gleaned from the guys while strolling around the farm. Until fairly recently it was a really sludgy and smelly sewage works with restricted access. It was the lack of access that was the major bugbear for many London birders and this issue ended up giving the place a bad name. All that has to changed. The site is now allied with the futuristic and nationally important Hackbridge Sustainable Suburb Project, a One Planet Living project, and the farm and environs are also a hive of recycling activity. Access, although still restricted, has been made easier for the general public with several organised open days throughout the year. The aim is that once the place is truly saved and the conservation work has been done, the doors will swing open not only for people but for the birds as well.

It was clear to me that the more well known Beddington Farm becomes among the birding fraternity, the more chance this area has of being the premier urban nature reserve it deserves to be.

Wormwood Scrubs, London

As a child birder I used to spend an inordinate amount of time in my bedroom. I was not playing with toys, reading comics or quietly contemplating world domination but instead pouring over field guides and writing lists in several notebooks. Yes, long before I was exposed to the ancient ways of birding I was instinctively compiling loads of lists. I kept a garden list containing some truly contentious species with gaps for others, such as Golden Eagle, that I was convinced would fly over my north London house at some point in the future. I had one listing extinct birds and even imaginary country bird lists for nations like Finland, whose borders I could just about recognise based on the distribution maps in my book – I even took a punt on the likely vagrants.

I gave up listing about fifteen years ago. I lie. Despite no longer indulging in a UK list, I still keep a world list, though don't ask me how many species I have seen. And I keep a list of the birds of Wormwood Scrubs, my beloved patch. The Scrubs list was born twenty years ago when I first set foot on its hallowed turf to make an inaugural late-summer visit. It was billed as a reconnaissance mission as I was more interested in looking at the available habitat and guessing at the types of species I thought might turn up. That day, I recorded a paltry twenty-two species but was encouraged by the good numbers of Song Thrushes and Linnets in the then sparse areas of scrub and woodland. A week later, after near daily lunchtime visits, I had clocked up an impressive list of passage migrants including two Pied Flycatchers, a Tree Pipit and a Common Redstart. My long love affair with the Scrubs began. It was a love affair that was to yield many surprises.

As already mentioned, my patch is a large area of parkland covering approximately 183 acres, which is even bigger than the nearby London Wetland Centre. A large portion is given over to playing fields but there is a thin strip of woodland that practically encircles the site dominated by Sycamore, birch, plane and oak. At the wilder western end of the Scrubs is a twenty-acre grassland, and the northern portion of my patch is bordered by a raised Gorse and Japanese Knotweed-clad embankment. Both of these segments have single-handedly turned up the best birds, including two Honey-buzzards, Osprey, Goshawk, Marsh Harrier, Wryneck, Ortolan and Little Buntings, Dartford Warbler, Quail, Great Grey Shrike and three Richard's Pipits. The grassland holds the closest breeding colony of Meadow Pipits to central London, while the embankment is much loved by good numbers of nesting Song Thrushes and Linnets.

My Scrubs list initially started as a daily log that evolved into a monthly then yearly total. For the first few years it was just good old me birding the site. However, after I began finding a few goodies more birders arrived and soon there was a small posse of 'Scrubbers', as we like to refer to ourselves. The list-making became communal and eventually we were setting ourselves annual targets. But listing at the Scrubs is by no means an easy task. It is an urban site surrounded by human habitation and frequented by people who like to wander everywhere. There is also no standing water, so immediately we are bereft of ducks, Coots and certain wader species apart from Snipe, a tiny number of which are regularly flushed in the winter. Indeed, the waders that we do get are invariably flyovers, as the Bar-tailed Godwits and Golden Plovers we have seen testify. We have even had years when a Mallard has not been recorded until March. I guess getting a good site year list is akin to a football season that runs from January to December. It is crucial that you pick up valuable points during the spring and autumn migration, when absolutely anything can turn up, to stand any chance of reaching your goal. It is during the migration season that we annually pick up difficult birds like Ring Ouzel and Common Redstart; the former is classically seen for a few brief minutes before spiriting away. A bad spring equates to a poor year-end total. It is as simple as that.

Every year we set ourselves a modest target of 100 species and to date we have never achieved it. The nearest we got to it was in 2010. By 28 December of that year we were on ninety-five and I remember desperately searching my snowy patch looking for anything to add to the total. Then I looked up to see a low-flying passage Woodlark heading west. It preceded a fairly large movement of Skylarks and was only our third record ever – ninety-six. I was still buzzing when an Egyptian Goose flew past. They nest fairly close to us in Hyde Park yet are rarely seen in our airspace – ninety-seven. I jubilantly sent a text to my girlfriend, who was in Thailand at the time. She sent me a text back saying '12 geese ahonking' – not quite correct for the Christmas carol. As I read it I heard honking. Thinking that I was dreaming, I looked up to witness a skein of thirty-two White-fronted

Geese pass low overhead heading east. We ended that year on ninety-eight species, but boy, what a way to end it.

York, Yorkshire

I have had a connection with York that spans nearly two decades. I used to have a longstanding girlfriend who hailed from the city; indeed, her parents still live there. Admittedly, the last time I visited this ancient walled city in north Yorkshire was in the mid-Nineties when I went to see a then little-known comedian called Eddie Izzard.

So it was high time that I returned to rediscover the birding attractions that I didn't consider back then. But I needed help. That help came in the shape of local birder Paul Brook, who drew up an itinerary that covered sites ranging from lush woodland to seemingly desolate urban wasteland, taking us from the outskirts to the heart of the city. The one thing that struck me about the city birding was the number of Tree Sparrows that we readily saw. Paul almost took them for granted, but for me seeing this nationally rare sparrow anywhere in the UK is always a very special moment. I was watching them alongside their more common cousins, the House Sparrow, almost as soon as we left my hotel to visit nearby Knavesmire and Knavesmire Wood. The first site was a small area of grassland that yielded countless butterflies, as well as several Reed Buntings and Whitethroats plopping into the long grass. The woods, managed by the Woodland Trust, flanked Knavesmire Racecourse – an area of short grass that must be a magnet during the winter for passing Golden Plovers and Lapwings.

We had a pleasant walk through part of the twelve-acre, sun-dappled Knavesmire Wood recording the usual woodland suspects ranging from Nuthatches to the ever-present Chaffinches. Despite being heavily visited by the local dog walkers and joggers, who I must add were all remarkably friendly, the woods seemed to hold a lot of promise, though they are sadly underwatched. The same could have been said for nearby Rawcliffe Country Park, another popular place for day-trippers, joggers and cyclists. The whole area, especially the contiguous Rawcliffe Meadows, was a haven for common warblers. Blackcaps, Chiffchaffs and Whitethroats frequently scolded us, while we searched in vain for Yellowhammers and soaring raptors of any description. At one point we were standing on a fairly busy path watching for movement in an adjacent hedgerow when I discovered a foraging Lesser Whitethroat. Paul was delighted because this handsome warbler is rather uncommon around York and his delight was amplified because it was his first one ever. I had found him a lifer.

The sun that had been beating down on us during the course of the morning was beginning to wane as we approached the riparian habitat at Rawcliffe Ings. Quite heavy clouds hung over us as we scanned the River Ouse for the Kingfisher that Paul knew hung out there. I felt quite confident that it was not about to rain because parties of Swifts were still darting above our heads. I'm not sure whether this is an old wives' tale, but I've heard that if it looks as though rain is on the way but the Swifts are still flying around then it means that it won't rain. If you look up and see no Swifts then you'd better run for cover. I have literally lived by this maxim for decades and I don't think that the Swifts have ever let me down. However, there is one major flaw with this method of predicting the weather: when the Swifts are out of the country sunning themselves in Africa, you can't leave home without a brolly.

Paul and I spent the rest of the day drifting through other potential birding sites in the city. Some, like Homestead Park, were municipal areas of green planted with stands of ornamental trees. We watched a Sparrowhawk, our only bird of prey of the day, soaring over the tops of the trees in this park and sending the local Goldfinches and Greenfinches into sheer panic. We also heard Goldcrests calling from a small stand of conifers, no doubt a good place to look during the winter months for a secretly residing Firecrest. York Cemetery looked good as it had all the necessary ingredients, such as wooded areas and plenty of scrub to attract potentially interesting migrants. The stretch of the Ouse opposite the Terry's chocolate factory produced a solitary Lapwing, singing Yellowhammers and a small colony of Sand Martins nesting in the riverbank.

Perhaps the most interesting find of the day was in an area not on the York ornithological map. We were driving near the centre of the city by an industrial estate when I noticed a large group of gulls resting on concrete waste ground. We stopped the car and as we approached the fence we noted that they were Herring Gulls. The birds were edgy and began to warily walk away from us, nervously watching us from over their shoulders. Suddenly, they took off in unison leaving one individual behind in their wake. This gull looked superficially like a Herring Gull but seemed smaller. We were confounded because neither of us had a scope and the bird was in a heat haze. Perplexed, I scanned the rest of the thistle-strewn wasteland and discovered three Little Ringed Plovers quietly standing in a group. Although synonymous with this kind of habitat we certainly didn't expect to find those birds. The identity of our mystery gull was never deciphered, but yet again we proved the point that you don't have to be anywhere special to see interesting birds.

Europe

Alentejo Region, Portugal

Since visiting Lisbon a few years ago I have found myself coming back to Portugal again and again. The thing that keeps luring me back apart from the Mediterranean climate is the rich diversity of birds to be found within the country's relatively small land mass. Most birders make a beeline for the fabled Algarve. Indeed, ask most people and they will tell you that Portugal is the Algarve. However, what they may not know is that there is a swathe of land that separates Lisbon in the north from the Algarve. This is the Alentejo region. It's not only an area brimming with wildlife, but has great food, wine and culture on tap. When I originally explored the region I didn't know what birds to expect. I too believed that Black-winged Stilts, Purple Gallinules and Greater Flamingos were exclusive to the Algarve, so I was certainly in for a surprise. Not wanting to sound like an advert for the tourist board, I think it's high time that I explain why I like this part of Portugal so much.

To do this I need to take you on a whistle-stop road trip. Our journey starts in the south-east of the region in Mértola, which is in the Guadiana Valley Natural Park close to the Spanish border and the Algarve region. I say Mértola but in reality I was birding a site called Mina de São Domingos on the edge of the town. It was a very spooky-looking disused mine with relict machinery standing idle in an eerie Martiancoloured landscape. Although it really did look like a set from a science-fiction film there were birds galore within its alien environment. It is a great place for seeing vultures and both Golden and Spanish Imperial Eagles, as well as a slew of passerines such as Rufous Bush Robin, Black-eared Wheatear and Thekla Lark. It was also here that João Jara, my guide, took me to a White-rumped Swift nest site that was secreted within the inner wall of one of the deserted structures. This species is a very rare breeder in Portugal with perhaps no more than twelve territories in the whole country. Because it was autumn we were safe in the knowledge that we were not disturbing any nesting attempts as the owners had already raised their brood and had long gone.

A little further up the road, well around thirty-seven miles as the crow flies, was Castro Verde. It is a vast area of rolling steppe land covering some 300 square miles. The Special Protection Area holds good numbers of speciality species – the kind of birds that are etched in blood on most birders' wanted lists. I was treated to lots of Calandra Larks whose distant flocks reminded me of waders clumping together, alternately showing their mantles and then their undersides. Their dark underwings strongly reminded me of Green Sandpipers, helping to further the shorebird illusion. We saw a few Black-bellied Sandgrouse and quite large flocks of Great Bustards on the horizon, and I learnt that it was one of the best places in Portugal to see Spanish Imperial Eagles or Iberian Imperial Eagles, as I am sure the Portuguese would prefer to prefix them.

The final fantastic destination is the Sado Estuary in Setúbal. At some 56,000 acres, the nature reserve, which is locally known as Reserva Natural do Estuário do Sado, is massive. It is surrounded by reed beds, paddy fields and in the south-west by a coastal sandbar. Despite the north-western border butting onto a quite densely populated area, it is still a birders' paradise. White Storks abound, nesting on nearby church towers and rooftops. Alcácer do Sal, a picturesque town at the

south-eastern edge of the estuary, prides itself as being the
Stork Capital. I popped in there for a cuppa and sure enough
I noticed that the stork population looked healthy. During
the summer scarcer Black Storks sometimes join the White
Storks in the fields. Truckloads of herons and egrets can be
found, as well as stacks of Glossy Ibises and Greater Flamingos.
The fields are policed by numerous Marsh Harriers, marauding
Peregrines and hovering Black-shouldered Kites, while on
the estuary itself Ospreys fish. Waders are also well represented
with all the species that you would expect to find on a
comparable British estuarine habitat, but with the added
bonus of Kentish Plovers, Black-winged Stilts and more
Avocets than you can shake a camera lens at.

The Alentejo region has to be Portugal's hidden secret and
it is so easy to reach from the UK as it is literally only a couple
of hours away. Birding in this region will definitely result in a
'phat' list (to use Ladbroke Grove street parlance) and a great
many memories. I personally will never forget an afternoon
that I spent quietly sitting by Lagoa de Melides, near Grândola,
further to the south of the Sado Estuary. I was scanning the
lagoon for Black-necked Grebes when at least 1,000 Glossy
Ibises suddenly flew low over my head in a straggling flock on
their way to another feeding area. They were totally unaware
of me as they passed over low enough for the swishing sounds
of their wings to be audible. It was a heart-stopping moment
and one that I felt privileged to have experienced.

Amsterdam, Netherlands

Generally speaking, the people that live in Amsterdam see
only two types of bird: Siskins and floating Siskins. Small

birds are Siskins and anything else are floating ones. I racked my brain trying to work out the logic of that statement as I embarked on my urban birding tour. I was struggling to keep up with my Dutch host and was peddling furiously through the streets of this famous city pondering: 'Why floating Siskins?'

Bradley Wiggins would not have been proud of me as I puffed and strained, wobbling dangerously as I tried to keep my balance while pretending to coolly look out for birds. Clearly, everyone in Amsterdam must have been born astride a Brompton. The city is the nerve centre of a very active birding scene in the Netherlands. Amsterdam seemed like a chilled and relaxed place, but was the birding as cool? My guide, Jip Louwe Kooijmans, was no slouch. As soon as we had met at the airport he was compiling a list of species that we were recording on the journey through to his BirdLife Netherlands office, where he heads the urban bird research team. The usual common suspects like Jackdaw immediately made it onto the list, but the best bird of the journey was the Great Egret that he spotted stalking in a ditch as we dashed past at high speed on a train.

Jip's office was situated in quite a leafy bit of town. The back gardens were big and wooded, and the general neighbourhood urbane and distinctly free from the city hubbub. Jays flopped from tree to tree whilst the expected Robins, Blue Tits and Collared Doves were all doing their daily thing. Cocking his ear, Jip thought that he heard a Crested Tit call from a particularly lushly vegetated garden. It is a bird that he fairly frequently encountered on his lunchtime walks around the neighbourhood. Back at his office Jip proudly showed me the BirdLife garden, which was littered with feeders. He told me of the Hawfinches and Marsh Tits that occasionally rubbed alulae with the more commonplace House Sparrows.

Where did the bikes come into the story? I hear you ask. Well, the following morning Jip had planned a looped cycle ride around one of the 'green fingers', as he termed them, to visit a few birding sites. It all sounded good to me until I straddled my bike. We had started in the eastern harbour district that was formally a fully functional port. Today it is a trendy area with stacks of waterside housing. Interspersed among the residences were occasional brownfield sites that sometimes hosted White Wagtails, and a Snipe or two, as well as the more obvious gulls and Carrion Crows. Flevopark was the entry point of our 'green finger' cycle. An old park with woods and a lake, it was designed by one of the first Dutch nature conservationists for city people to enjoy the pleasures of nature. It is a site with an ornithological pedigree, as earlier in the spring a River Warbler spent three weeks singing from the tops of the headstones in the overgrown cemetery that forms part of the park. For us, Nuthatches, tit flocks and Chiffchaffs were the order of the day, though I did manage to glimpse a Lesser Spotted Woodpecker bouncing away through the tops of the trees. Unbeknown to me, I had just seen a city rarity.

The bike ride to the nearby Science Park in the midst of some new university buildings was particularly gruelling as not only was I already feeling pretty saddle sore, but I also had to negotiate both peddle-powered and motorised traffic – on the wrong side of the road. The area was essentially a very overgrown building site that will probably have buildings plonked all over it by next year. Interestingly, all construction sites in the Netherlands are on standing water. This boggy ground encourages the growth of reeds and other riparian flora. Wading through the man-sized hogweed, we flushed a surprising twenty-plus Snipe and encountered several Reed Buntings. In the summer months a pair of Bluethroats held territory, making them one of the closest breeders to the city centre. Over the past two years Jip has also recorded nesting Black Redstarts and watched Spoonbills foraging on the man-made muddy pools. It just shows what life can exist even on the most temporary of urban landscapes.

My favourite spot was Diemerpark, not least because it was there that I discovered that my bike had gears. On paper the site didn't sound very exciting as it's surrounded by extensive housing estates and is perhaps the most polluted area in the whole country. But it was. There was a good variety of habitats, including grassland, woodland and a harbour. The birdlife reflected the range of landscapes. It is one of the most heavily birded areas of the city and as a result good birds have been uncovered, such as a recent Woodchat Shrike. Common Buzzards and Red-crested Pochards breed and on our visit we successfully tracked a recently reported Cetti's Warbler. Did you know that in the Netherlands the song of this distinctive warbler is transcribed as 'Hey! Get out of the car, bastard!'?

I had a great day birding. Dutch experience over, I got off my bike, said my goodbyes and hobbled bow-legged to the airport and into the sunset.

The Azores

Every now and again, even an urban birder like myself needs to get out of the city sprawl to grab a breather somewhere as far away from a streetlight as possible. I grabbed an opportunity to do just that one October when I journeyed to the Azores in the company of *Bird Watching* magazine's Stuart Winter and my photographer Russell Spencer. The Azores is a cluster of nine volcanic islands set on the extremity of the Western Palearctic in the middle of the north Atlantic. They are revered as *the* place to be in the autumn by many European birders. As you are probably aware, the islands are a red-hot spot for American vagrant birds. On any given day during the migration period falls of individual birds of many New World species are commonplace, hence the attraction to birders keen on adding to their lists birds that are normally ultra-rare on the European mainland.

We touched down on the main island, São Miguel, and were met by our guide, Gerby Michielsen, a nervy and hyperactive Dutchman who had settled on the islands a decade earlier. The Gerbster, as we quickly nicknamed him, immediately set about showing us many of the spots within Ponta Delgada, the main town on the island and, indeed, the archipelago. Soon we had clocked up a couple of Pied-billed Grebes and a Blue-winged Teal. We also connected with most of the few resident birds available on the Azores. I noticed that the Chaffinches were darker than our birds and seemed to flutter more, the Blackbirds had strange weaker calls, Grey Wagtails occupied every wagtail niche and Atlantic Canaries were everywhere.

Now, although I like a rare bird every now and again, I hadn't come to the islands for the sole purpose of ogling waifs and strays. This was a fact that became patently obvious when we flew to the rarity hunter's paradise, Corvo, the smallest and northernmost island in the group. The afternoon we spent there resulted in several more Blue-winged Teals, a Double-crested Cormorant and a ridiculously close White-rumped Sandpiper. I didn't take to Corvo. I found the idea of pounding well-worn paths just looking for rarities a turn-off. I could understand and get involved with looking for oddities on islands off the coastlines of Britain and Europe, but in the middle of the Atlantic? Wouldn't it be better to just travel a little further west and watch them flying around in situ with tons of compatriots in the 'land of the free'?

We left Corvo by boat over choppy seas and under grey skies heading for nearby Flores. Hundreds of Cory's

Shearwaters entertained us as they wheeled and sheared around the boat, while ubiquitous Azorean Yellow-legged Gulls welcomed us at the harbour. Far bigger than Corvo, Flores had a lush appearance and, crucially for me, was underwatched according to the Gerbster as most birders visiting the Azores just make a beeline for Corvo. Filled with excitement, we got swept up in a twitching fever and rushed over to a local football pitch to see a recently discovered Upland Sandpiper. To be honest, this weird-looking relative of the Curlew was a bird that I had always wanted to see and once we had refound it I spent ages studying its behaviour and its quirky look.

The next four days involved watching a couple of patches with the occasional forays to other potentially interesting sites. Stuart found a Black-and-white Warbler whilst investigating an area of gardens, and Russell discovered a Tennessee Warbler near the coast. Not to be outdone, the Gerbster found a Grey Catbird and I stumbled across a dowitcher that we saw all too briefly to assign specifically. We shared a surreal moment when a flock of four White-rumped Sandpipers that we found on our rounds quickly became a threesome after one of their number was swiftly dispatched by a buffy-looking juvenile Peregrine, which we momentarily thought was a Hobby. We figured that it might have emanated from the cold Canadian tundra.

I really enjoyed being on Flores and hanging out with the guys. The island was truly beautiful and satisfied my desire to be on a patch where I could find and enjoy my own birds. We ended up staying an additional week thanks to an almighty offshore tropical storm in the wake of Hurricane Otto, which resulted in flights being cancelled. Great, you may say, but we needed to get home. Worse still, one morning I awoke to a text message that resulted in tears of anxiety and depression. A Great Grey Shrike had been discovered on my beloved Wormwood Scrubs – a bird whose appearance I had predicted – and I was stuck on an Azorean island thousands of miles away. Ironically, had I been in the UK I wouldn't have been able to twitch it anyway because of a commitment I had made to be up north delivering a talk to the Preston Bird Watching and Natural History Society. To add insult to injury, when we finally got a flight back to San Miguel, we managed to dip out on the Azores Bullfinch, the Azorean signature bird, while searching for it in the last hours of daylight and in chucking rain.

The Azores may not appeal to every birder but there is far more to these fascinating islands than just chasing rarities. Having said that, it's hard not to ignore the potential of finding something interesting and the chances are that you will. If you are up for that then you will have a great time.

Belgrade, Serbia

Love it or hate it, Facebook and other social networking sites have done a lot to break down international barriers. In an instant you can find yourself communicating with people in far-flung places around the globe – people that you would never have come across in real life. As I glance through the list of my Facebook birding friends, at least seventy-five per cent are still unknown to me, a few others I have got to know in a virtual sense and the tiny remainder consists of people I have actually met. One person that came through all three stages was Serbian fellow Facebooker Dragan Simic. Inspired by my Tower 42 bird-study project in the heart of London, he sent me a message out of the blue inviting me to Belgrade to help set up a similar project and to indulge in some urban birding. I paused for thought as the city was certainly on my radar. I didn't really know

too much about the birds I could expect to see but I had
heard a fable about an amazing winter city roost of Long-
eared Owls that was numbered in hundreds. It seemed to
have the hallmarks of a great urban birding venue. I had to
investigate.

Within two months of that fateful Internet conversation I
was on a plane heading for the White City if you're Serbian,
or Belgrade to the rest of us. Situated on the confluence of
the Danube and Sava Rivers where the Pannonian Plain
meets the Balkans, its location sounded magical. Having left
a decidedly chilly autumnal London a couple of hours
previously the first thing that struck me when I got off the
plane was the heat. It was surprisingly hot. The woolly hat
and long coat that I was wearing had Dragan laughing when
he met me. Serves me right for not checking the weather
forecasts.

Dragan was an energetic and knowledgeable man who
had met with an unfortunate sports accident some years ago
that left him permanently on crutches. Despite his disabilities,
he was still able to drive around and never once complained
about his predicament even though it meant that he couldn't
walk very far. I marvelled at his courage. Our birding mission
got under way as soon as we started the drive through
the busy charismatic streets of Old Belgrade. Kalemegdan
was our first stop. It is a fairly wooded park that contained
a Turkish fortress directly overlooking the confluence of
the rivers.

Despite being late in the afternoon and sharing the park
with everyone, his wife and their dog, we did manage to see
several migrant Spotted Flycatchers hawking insects and
glimpsed a few Blackcaps too. At ground level there was a
variety of interesting animals, including some nice-looking
lizards and plentiful Swallowtails, a butterfly I only rarely see
in Norfolk. Kalemegdan is certainly underwatched and I
could see this site's potential for the early-rising birder.
Looking east across the confluence lay the large and heavily

wooded Great War Island that was once the home of a fairly large egret colony. Thankfully, these birds had relocated themselves to a nearby riparian wood and the island now boasts a winter roost of over 2,500 Pygmy Cormorants – a target bird for many British birders. Plus, there are regular sightings of urban White-tailed Eagles there. An incongruous thought, surely?

Dragan had thoughtfully booked a short boat trip around the island. The evening light began to wane as we sat on the rickety craft navigating the circumference of the island. Black-headed Gulls abounded, with the larger gulls being equally assigned to Yellow-legged, Caspian or 'another big gull'. A solitary Common Buzzard drifted through the treetops to roost in the interior of the island just after we had discovered a couple of locally scarce Little Gulls that had secreted themselves among the throngs of their Black-headed cousins feeding over the river. Several bats, probably Noctules, hunted high over the trees like large hirundines. This all sounds rather romantic and idyllic, but when I looked at the river I was shocked at the amount of filth and garbage floating downstream. There is obviously a lot of education needed in this city to teach its populace the importance of conserving its environment.

The next beautiful clear morning saw me and thirty birders on top of USCE Tower for the inaugural session of watching for visible migration. After being interviewed by the press and a couple of TV stations we soon found ourselves on the roof of this well-placed office block over 400 feet above the city with unparalleled views of the confluence and the island. From this lofty perch the city looked pleasingly green and wooded. This newly opened observation point promised some great birds in future watches if the previous sightings of some good raptors are anything to go by. We were unlucky and didn't see much aside from the obligatory Caspian and Yellow-legged Gulls, but it was great being up there. The other observers were also suitably thrilled. This building had been out of bounds to the general public, so they relished the opportunity to spot the areas where their mums lived.

I really enjoyed my impromptu Belgrade visit watching the default urban birds like Rooks, which seemed to be everywhere, a few Hooded Crows, eastern race Jackdaws and plenty of chirpy House Sparrows. Had I visited during summer I could have enjoyed delights like Golden Orioles and even Bee-eaters that breed in the city. Dragan told me about a flock of more than 500 migrating Common Cranes that had settled between some inner-city tower blocks.

And as for the roosting multitudes of owls, they were there too. I will have to come back one day to check them out. Belgrade clearly has a lot to offer.

Bratislava, Slovakia

I do love the idea of cheap air travel – carbon-footprint issues aside for one moment. I mean, who would have thought ten years ago that we would be able fly to amazing European destinations like Budapest or Prague for as little as a pound each way. In reality you always end up paying much more than what the budget airlines advertise, but it still works out to be pretty cheap.

What people tend not to do a great deal, though, is go birding when they're away on their weekend city breaks. I think that finding a city patch abroad can be a very exciting prospect as there is often no information available as to the species likely to be encountered. You truly become a pioneer. The thought of heading off to a foreign city always stimulates my desire to eke out some urban birdlife.

I'm not a complete glutton for punishment because I often try to build in some time to head out of the metropolis I'm staying in to visit a 'proper' birding site. A few years ago I spent a weekend in Barcelona with my then girlfriend to work on our floundering relationship while taking in some culture. Sensing that the end was near, I ditched the idea of a reconciliation and dragged her kicking and screaming to the world-famous Ebro Delta about sixty miles to the south, for a birding 'fix'. I had a great time watching great birds like Little Bittern, Short-toed Lark and Black-eared Wheatear just outside Tarragona. Of course, my soon to be ex-girlfriend was not wildly impressed, especially when our hire car got a puncture whilst I skidded on a dusty track trying to avoid a tractor that was heading towards us at full tilt.

Recently, I had an invitation from a Slovakian friend of mine to spend a long weekend in Bratislava. I didn't need much persuasion. A quick Internet search got me a cheap flight, but further investigation resulted in zilch information on the birding sites in the city, let alone the birds. I was on my own. Well, even in the worst-case scenario I would have been able to immerse myself in Slovakian culture, marvel at splendid architecture and laugh at drunken Brits on their stag nights.

Nonetheless, knowing that Slovakia is generally a great birding location and excited by the prospect of some potentially interesting urban birding, I jumped on a plane and arrived in heat-soaked Bratislava a few hours later. After checking in at my hotel smack in the middle of the gorgeous Old Town and yards from the mighty Danube, I picked up a cheap and nasty street map given to me by the hotel and started making plans. I quickly discovered that there were areas of 'green' on the other side of the river. That was all the information I needed. So, the following morning I arose at sunrise and five minutes later I was walking over one of the plentiful bridges that span the Danube to investigate.

The first area I came to was a municipal park populated by tall old isolated trees. I immediately clocked up Green and Great Spotted Woodpeckers, Blackbirds and several Black Redstarts – but little else. On the nearby river were a few Black-headed Gulls and loads of hawking Swifts and House Martins. I continued walking east along the river until I came across another bridge with a curious flying saucer structure on top of it. I later found out that this was Nový Most (New Bridge) and the 'UFO' bit was indeed a restaurant. Apparently, during the communist days Bratislavans were not allowed to see the view from this vantage point for fear that they would be tempted by the sight of nearby capitalist Austria.

It was at the bridge that I stumbled into the other ambiguous green patch on my near-useless map, Sad Janka Král'a. Named after a Slovak poet it is a heavily wooded park that is apparently one of the oldest in eastern Europe. I later discovered that despite this site's apparent richness in wildlife, the developers are hovering with their beady eyes peeled and chainsaws revved. I immediately saw the potential here for some reasonable birding and for the remaining three days of my stay I made the area my Brat Patch. My dedication was quickly rewarded with plentiful Hawfinches, a bird so difficult to catch up with in Britain, Blackcaps, Spotted Flycatchers, Hooded Crows and even more Black Redstarts.

An immature Hobby drifted over low one morning being mobbed by squadrons of House Martins, which nest on the undercarriage of Nový Most. The ornithological delights of my patch were the daily fifteen or so Golden Orioles that I saw flying around the canopies, singing and travelling from one clump of woodland to another. A family of Collared Flycatchers was also a delight to see despite being a nightmare to decipher from the very similar Pied Flycatcher.

Perhaps my favourite sighting was of a pair of Red-backed Shrikes feeding what seemed to be just one fledgling in a small clearing on the edge of the woods. I watched the stunning male for nearly half an hour as it hunted large burrowing wasps on the cycle track that bypassed the clearing. Interestingly, during my prolonged study of the birds I discovered that the family unit consisted of two males and a female.

Before I knew it, I was back on the plane heading back to England with memories of a fantastic city with a birdlife to match. The moral of my tale is quite simple. Don't be afraid to explore a new city to uncover its avian secrets. Once you have found an area to study always have faith in your chosen patch even if, like with my Brat Patch, it is popular with cyclists and dog walkers. And finally, get a decent map.

Brussels, Belgium

Every now and again I like to give myself a real urban birding challenge by visiting metropolises that are little known for their birding potential. On paper, there is perhaps no European city that has less perceived birding credentials than Brussels. It is the seat of the European Union and conjures images of officialdom, boringness and a sea of grey concrete buildings, some historic and others just grey, but all occupied by people in suits. Oh, and not a bird in sight. Ah, a perfect venue for the Urban Birder to explore.

After a short couple of hours on the Eurostar I stepped off in Brussels to be greeted by expat polyglot Stephen Boddington. A citizen of the city for over fourteen years, he had promised to show me the hidden avian delights that were on offer. Stephen was a very interesting character. Aside from his linguistic skills, which included fluent French,

Dutch and Esperanto, essential for life in Brussels, he was also an adept birder who rejoiced in studying raptor migration across the world. He then unashamedly revealed himself to be a major Eurovision Song Contest fan-club member, causing me to involuntarily spit the orange juice that I was drinking into his face. I guess all birders have their quirks.

Strolling down the hill from the upper city to the lower city in central Brussels we passed a few hedges where on previous nights Stephen had flushed migrant Woodcocks. Indeed, you never would have assumed that Brussels is a good venue for migration watching, but passage clearly occurs through the city. There is even a keen urban birder who watches diurnal migrants from the lofty heights of the Town Hall's tower in the architecturally beautiful Grand Place, the main touristy square in the city. Among the migrants that he has noticed passing through have been drifting Honey-buzzards and Marsh Harriers, along with good numbers of Common Cranes. The Grand Place is also the spot to hear Black Redstarts in spring singing above the thronging crowds.

The early-spring sun shone brightly over us as we walked the historic streets towards Brussels Park, our first proper venue smack in the centre of town. We were greeted at the gate by an Egyptian Goose standing regally on guard atop a statue of a cherub. The park itself was rectangular and though there were a few stands of trees it was basically a manicured lawn interspersed by paths. We viewed the nearby private gardens of the king's palace through gaps in the walls. The bushes along the walls often attract Firecrests, though we could not muster anymore than a handful of Magpies and a pair of Jays.

As our plan was to travel around the city via public transport we duly boarded one of the plentiful underground trains and resurfaced near Woluwe Park, a recently discovered birding spot on the eastern side of the city. Its seventy-one acres of woods, parkland and a lake make it one of the largest green spaces in the city, and it enjoys a reasonable exchange of woodland birds from the nearby forests that skirt Brussels. The story goes that just a few years ago somebody noticed that Middle Spotted Woodpeckers were fairly reliably seen here and before long it was announced as the best urban venue for woodpeckers. During our visit we had sightings of several courting pairs of Great Spotted Woodpeckers, some decent views of Middle Spots and a female Green Woodpecker entertained us. We also heard a brief resonant snatch of drumming that could only have been orchestrated by a Black Woodpecker. Short-toed Treecreepers were not in short supply and neither were the Nuthatches. Tufty Red Squirrels were also running riot. It is here that I discovered the true extent of Brussels' introduced avian exotica which included Rose-ringed and Alexandrine Parakeets, tons of E-Geese (Egyptian Geese to you), Mandarin Ducks and even a pair of Wood Duck sitting in a tree. To complete this feral menagerie Siberian Chipmunks scampered on the woodland floor. We left the area after watching the more natural sight of displaying Stock Doves in numbers that surprised Stephen as he had normally regarded them as scarce.

While waiting to catch a tram to our next destination, we did the classic urban-birding move and looked up. Shortly, we were rewarded with a mini-raptor passage. The initial single thermalling Common Buzzard was joined by three more of its brethren, a pair of Sparrowhawks and, best of all, a Goshawk that sailed directly over us. We were still elated when we arrived at the forested south-eastern edge of the city at Rouge Cloître. This nature reserve consists of a group of wood-fringed ponds that join up with the outlying forest. More woodland birds awaited us but we missed the wintering Red-crested Pochards, which had moved on.

My day in Brussels drew to a close and we found ourselves back in the city centre watching a screeching male Peregrine with prey entering the spire of the cathedral near Central

Station. Its mate was no doubt sitting on eggs. Brussels was a far better urban-birding venue than I had ever imagined. The only spoiler is the ongoing feud between its Dutch-and French-speaking citizens. This mutual disdain has even permeated into the birding community, with both sides having their own birding websites with largely no crossover. Why can't birds of a feather stick together?

Budapest, Hungary

I feel like I'm one of the luckiest people on two legs. Professionally, I am doing what I love and I have amazing opportunities to travel the world in the pursuit of birds and other wildlife. It's a dream. But I also feel guilty visiting some of the places I do because I never have any time to indulge in the fabulous architecture, fantastic art and culture that the cities that I write about have to offer. I'm too busy looking for birds.

The guilt I felt was trebled when I visited Budapest recently. I was invited over by the Hungarian National Tourist Office, who along with their partners did everything but roll out the red carpet to make my stay as comfortable as possible. I stayed in a fantastic health and spa hotel on Margit-sziget (Margaret Island), a wooded island in the Danube between Buda and Pest. Needless to say, I didn't get to sample the saunas, massages and cosmetic surgery that were on offer to me.

My first experience of birding in Budapest came shortly after my arrival when I met my guide, the renowned Gerard Gorman, Mr Eastern Europe himself, within the parkland grounds of the hotel. We planned the itinerary as we walked around listening to singing Blackcaps, Nightingales and Golden Orioles. Gerard is a straight-talking Liverpudlian who settled in Hungary over eighteen years ago and is probably the best person to be with when birding in eastern Europe. He has an overriding passion for woodpeckers and has even written a book about them. I knew that I would be in safe hands when it came to seeing the last of the European peckers that still eluded me.

I was happy when Gerard asked me to join him the following day birding outside the city in the company of an elderly American lady who had booked him to find her some lifers. It meant that I would be away from the temptation to stare in admiration at some of the gorgeous buildings that I had previously only glimpsed. We picked up Joan – all five feet of her – by the Danube in the shadow of a pretty bridge in the centre of town and headed off to the Kiskunság region, some fifty minutes' drive south-east of the city. Joan, eighty, introduced herself as 'Joan the Snack Queen'. She wasn't lying. I felt like the runt chick in a nest in the back of Gerard's car with Joan sitting in the front busily stuffing morsels down Gerard's throat from her ample rucksack. The occasional biscuit was thrown my way when his mouth was too full.

She was an absolute delight to be with and I have never met an octogenarian who had the energy and drive that she had. With more than 6,700 species on her world list and at least one member of every bird family known to science under her belt, she had a very definite idea as to what species she wanted to add to her list. The Kiskunság region was a gorgeous rustic idyll that encompassed grassland, farmland, woodland and fish ponds. Indeed, sections of the area have been designated as a national park. It was also bird-rich, so Joan soon added the species she desired. Rollers and Bee-eaters, together with Red-backed Shrikes, Corn Buntings and Turtle Doves, were easily seen. We also saw goodies like Lesser Grey Shrike, Tawny Pipit, Great Reed Warbler and

Penduline Tit, plus waterbirds like White Stork, Spoonbill, Little Bittern and Purple Heron. I saw my first undisputed Caspian Gull and again, frustratingly, heard my two bogey birds: Quail and Savi's Warbler. Our day out in the country was capped by an Otter, which ran across the road in front of us while we were stationary in the car at the fish ponds area.

On the way back we dropped into Csepel Island, an urban park on an island in the Danube, to look for Syrian Woodpecker – a tick for both the Snack Queen and I. Within minutes we were celebrating, watching a pair coming to a nest hole to feed a large juvenile with its head poking out of the hole. Gerard took us through the subtle differences between this species and the very similar-looking Great Spotted. They have salmon-pink undertail coverts instead of scarlet ones, a white face without the black border, a thinner bill and a subtly different voice. But most interestingly, the Syrian is truly an urban bird, hardly ever being found in the woodland that surrounds Budapest, whereas the Great Spot is quite the opposite.

Gerard and I spent the next day visiting the various noteworthy urban sites within the city. I started my education in his local patch at Normafa, the woods above the city in the Buda Hills. We saw scampering Red Squirrels, listened to and occasionally glimpsed wailing Black Woodpeckers, found a garrulous family party of Middle Spotted Woodpeckers and stopped to watch a pair of Collared Flycatchers attending to their hidden nestlings. Later, we were in Farkasréti Cemetery in the 12th district of the city. This was a typical Budapest necropolis that attracted the usual array of Blackbirds, Blackcaps and other woodland species. We were lucky enough to locate a grey-phased Tawny Owl as it sat in a graveside willow being furiously mobbed by Blackbirds.

I had an absolutely brilliant time in Budapest and when Gerard and I parted company, I went for a final stroll in the grounds of the hotel to reflect on my experiences while soaking in the singing choir of Nightingales and Blackcaps.

Then I had an interesting thought. There are 3,400 cities with populations of more than 100,000 in the world today. If I visited one every weekend it would take me about sixty-six years. Maybe then when I'm 100 I will have the time to visit them again, this time purely to go sightseeing.

Helsinki, Finland

I was warned before going to Helsinki not to be duped into staying up all night lulled by the perpetual daylight of the Finnish midsummer. I know what I'm like – I'm a sucker. If there's light, especially if there is sunshine, then there will be birds to be found. Well, that's what I like to believe. It's a notion that is ingrained into my psyche. This behaviour stems back to when I was a hugely inquisitive kid hell-bent on learning everything possible about birds. I always had this sense that I would be missing something if I wasn't at my patch to greet the sun as it rose. There was method to this madness, though, as I believe that there is an element of truth to the old adage 'the early bird catches the worm', especially in urban areas. I can recount the many occasions during which had I not been out birding at the crack of dawn I would not have discovered the Short-eared Owl that flapped lazily over, or flushed the Snipe that had spent the previous night gorging in the grassland of an unlikely urban paradise.

So I guess I am a prime candidate for falling under the spell of the alluring everlasting northern summer light.

The warning I received was a wise one because a week spent partially in Helsinki and eventually further north in the taiga forests in the Kuhmo region near the Finnish-Russian border was testing. I felt like Al Pacino's character in that film *Insomnia* where he went up to Alaska to investigate a crime and ended up craving a bit of night-time darkness.

But back to the main reason for my trip to Finland: to eke out the birdlife of Helsinki, a small but pretty cool city. I was no stranger to the city as I had once been there on business for a whistle-stop summer visit, though on that occasion my birding was extremely limited. Back then the couple of hours that I managed in a small park close to my hotel resulted in copious Fieldfares, a few roving Blackbirds, Chiffchaffs, Great and Blue Tits, a couple of Willow Warblers and an obliging Pied Flycatcher. I was also lucky enough to be taken by my hosts to their summerhouse on one of the thousands of islands that litter the Helsinki Archipelago some fifty miles outside the capital. I'm sure that this island sanctuary had a name, but it looked identical to all the other islands dotted around close by. On the way, I managed to see a pair of impressive Ospreys at their nest as we sped past another nameless island by speedboat. Aside from being shown how to smoke trout in a rusty-looking tin box, bake blueberry pie using locally plucked blueberries and hearing about the intricacies of naked axe-throwing (a local pastime I'm told), I flushed a Goshawk while on a short post-lunch walk around the island.

On my next visit three things immediately struck me as I journeyed from the airport to my quayside hotel in the blazing heat that would not have been out of place in southern Spain. First, Common Gulls were everywhere – strolling in the streets, and sitting under trees in the parks and by the waterfront all around Helsinki. They were the default larid. I always thought of these gulls as gentle looking, inoffensive birds that spent most of their time lounging on football pitches and trotting on the spot to bring up earthworms. That illusion was shattered for ever when I observed a noisy group chasing a congener who had just nicked the Finnish version of a Mr Whippy from the hand of an unfortunate tourist. They had obviously been mixing with the Herring Gulls from Brighton.

The other thing I noticed was the extraordinary number of Barnacle Geese in the city. They were as numerous as the Canada Geese in London. My guide, Hannu Tammelin from BirdLife International Finland, told me that although a proportion of these birds emanated from escaped stock, there was a large number of genuinely wild birds too. I loved them. They were cute looking and quite wary in a tame sort of way, not allowing you to approach as close as you can to their larger Canadian cousins. The locals evidently didn't share my naive enthusiasm for them and many think of them as pests. Perhaps the most ubiquitous bird, though, was the Fieldfare, the typical thrush of the many northern European cities that I have visited. It was even more abundant than Blackbirds are in comparable habitat in Britain.

There is much more to Helsinki than those easily seen birds. However, this was something I had to wait until 3 a.m to experience, as my first job was to ensconce myself in my hotel room to finish writing an article for *BBC Wildlife Magazine* that should have been with them the previous week. I had twelve hours, so there was time – or so I thought. It was eventually 3.15 a.m. when I finally pressed the send button. I was shattered and seriously needed to sleep. Hannu had been waiting in reception for the past fifteen minutes and worse still, it was daylight. I couldn't have slept even if I had wanted to.

Hannu nervously greeted me as I sleepily tipped out of the lift into the reception. He was a middle-aged man of a seemingly shy disposition. He led me outside while running his thoughts for the day's itinerary past me. Our first call would be at the nearby Torni Hotel not to partake in breakfast but to stand outside the hotel to stake out an urban Eagle Owl that was supposedly nesting on a nearby rooftop. He led me to a rather smart-looking black Mercedes. I was puzzled. Somehow, the expensive car didn't quite fit with the humble man that I was standing next to. It was then that I noticed that the car had a driver. Hannu saw my confusion and explained that the car was ours for the day at the expense of the tourist board, as he was a non-driver.

Our early-morning street stake-out resulted in zilch. I actually returned the following evening at 11 p.m., but this time like any good detective I snuck into the hotel and secreted myself on the roof terrace with a great vantage point over the rooftops and indeed the city. I must have cut a desperate figure, squashed up against a plate-glass wall peering over the edge with a vodka and orange in one hand and binoculars in the other, mobbed by a bunch of drunken singing Finns. Needless to say, I soon gave up that stake-out.

Our driver ferried us to a far more successful port of call – Iso Huopalahti, a large, serene, reed-fringed lake at the western edge of the city. It felt like late morning to me but it was still only 5 a.m. Walking around the water's edge, I listened to River Warblers, and glimpsed Blyth's Reed Warblers in the waterside bushes, as well as watching Thrush Nightingales out in the open foraging on the paths ahead of me like Robins without red breasts. The Blyth's Reed Warblers were at best flitting shadows in the bushes, although Hannu was totally confident about their identification. He was obviously completely up on his local birds. I didn't share his confidence as I had had no previous experience with this *Acrocephalus* and the views that I had were far from conclusive. My modus operandi is that if I can't recognise a species for myself, then it doesn't go on the list, regardless of the expert standing next to me. I have to see and believe with my own eyes.

Further west in Suomenoja, a lake overshadowed by a big factory, Hannu and I enjoyed the sight of up to 4,000 Black-headed Gulls with their recently fledged young. Some of the adults were airborne or clumsily sitting on the telegraph wires that criss-crossed the vista. I know that they are common birds in Britain but I still find them mesmerising to watch in such large congregations. Swimming on the lake itself along with some of the gulls were Mallards and a few beautiful Slavonian Grebes; some migrant Wood Sandpipers stalked around the muddy edges. On the adjacent derelict land were more of the ubiquitous Fieldfares, plus a lovely male Northern Wheatear and a Stonechat or two.

Vanhankaupunginlahti (try saying that after a few drinks) was next on the agenda, as Hannu wanted to take me on a twitch. This large lake has a big expanse of reed bed and is home to Marsh Harriers, several warbler species and many Reed Buntings that cavorted alongside the singing male Common Rosefinches, whose red plumage kicked out from the straw-coloured reeds. Redwings and Fieldfares were also evident, and a distant Kestrel added a bit of raptor spice.

However, our quarry inhabited the meadowland to the east near another large lake. We climbed a tower hide that overlooked the area and were soon observing a few Caspian Terns by the lake, which were accompanied by yet more Wood Sandpipers and some male Ruff that were still practically in full summer plumage. From our lofty position we were also eye to eye with a Stock Dove that was perched on top of a nearby dead tree. Hannu then found what we were looking for. In a fairly distant meadow was a family party of Citrine Wagtails. I watched with delight as the adults attended to their two fledglings. My attention was only diverted to quickly look at an Icterine Warbler that had burst into song in the trees directly behind us. The wagtails were truly unusual as they were well west of their normal breeding range.

The great thing about Finland is that people are really encouraged to get out and respect and enjoy nature, and Helsinki is a great place to start. There is a large number of islands to explore in the archipelagos. Some like Seurasaari are wooded and are supposed to be good for Black Woodpeckers despite being quite touristy. An early-morning cycle ride around this island resulted in the usual Helsinki suspects plus plenty of inquisitive Red Squirrels. They would run away, stop, then run back towards me as if to take a closer look. I was even lucky enough to discover an immature Long-eared Owl hunting and eventually settling on a low branch close to me.

My favourite island by far was Harakka. Literally a few minutes' ferry ride from the city, this island, whose name means 'Magpie', is an urban nature reserve, which is quite rich in flora and fauna. Previously a military base until 1989, it now hosts several visitor centres manned by some very friendly staff. There is a large breeding population of Common Gulls and scattered among them are a few Baltic Gulls – or the *fuscus* race of the Lesser Black-back if you are not a splitter. Northern Wheatears obviously breed as I found quite a few fledglings, and I heard loads of Willow Warblers and Lesser Whitethroats. Along Harakka's coastline were many Common Terns, Cormorants and Eiders. Walking around this island I could really see its potential as a migrant trap as one side faces out into the Baltic Sea. It's the sort of place that I would adopt as my local patch if I were a resident of Helsinki.

My time in the city had come to a close as I was due to head further north to observe bears and Wolverines. As I said earlier, Helsinki is a great city with loads of cultural stuff to do. Of course, as an urban birder the call of the wild was a far stronger draw and the prolonged daylight meant that sleep deprivation ruled. My advice when visiting Helsinki is to make sure you get yourself to bed at a reasonable time.

Kraków, Poland

My grasp of Polish is rudimentary despite having had a long-term Polish girlfriend many moons ago. During our relationship I only managed to learn how to say hello, thank you and ankle (don't ask). So, when I met my guide, who had come to pick me up from Kraków Airport, I proudly trotted off my entire Polish repertoire – an act that took all of five seconds. He smiled quizzically at me as he probably thought, 'Who is this idiot?'

He was a lovely man with a name I couldn't even begin to pronounce. I resorted to politely referring to him as 'hey' and 'er' for the first day and a half. His name was Przemysaw and I'm certain that even he had trouble pronouncing it. Eventually, after consulting his wife, I realised that Przemysaw sounded like Chemink (as in Temminck's). I rechristened him Chem from then on.

My first day in this architecturally beautiful city was spent walking practically from dawn to dusk. Chem was a big believer in leaving the car at home and getting involved in bipedal action. Sunrise saw us walking through the old town observing plentiful Fieldfares and Blackbirds hopping around on the lawns of Planty, the strip of parkland that encircles the core of the city. It was strange to see an active rookery in a city centre with at least twenty nests, as I always equate Rooks with the countryside. Jackdaws were plentiful, too, although these birds had particularly hoary napes thus having a distinctly different look from our British birds.

Our destination was Las Wolski, a large tract of forest in the western suburbs of the city. Chem was a lover of forests and his day job involved working in forest nature protection. Las Wolski covers an area that is a third of the total 3,700 acres of forest that remains within the city. I had a fantastic time wandering around listening to the multitudes of Chaffinches singing their subtly different-sounding songs and watching gangs of Hawfinches that seemed to be everywhere.

Chem's command of English was pretty good for a man who had never stepped foot in our green and pleasant land, but he was seriously struggling to understand me when I was excitedly shouting out the names of the birds in English. Unless the words hello, thank you and ankle were involved, there was no point in him even thinking about telling me what their equivalent names were in Polish. We were both frustrated until I had a eureka moment. '*Dryocopus martius*' I exclaimed in pigeon Latin. A broad smile of recognition came across his face as he corrected me fluently. We had just seen a Black Woodpecker. All those pre-teen years spent learning the Latin names for every single bird in my Heinzel, Fitter and Praslow's *The Birds of Britain and Europe with North Africa and the Middle East* had not been in vain.

Thus started a really fruitful birding relationship as we clocked up some of the woodland speciality species. I was delighted that I was able to surprise Chem on his own home turf by finding two locally rare species – a spanking male Collared Flycatcher and a handsome male Middle Spotted Woodpecker, which was also a lifer for me. Later, we went for a walk along the natural banks of the nearby River Vistula, where aside from Chem seeing his first Common Tern and Swallow for the year, we also saw a couple of frog species, a few Grass Snakes and evidence of tree felling by the local population of European Beavers. Inner city beavers, who would have thought it?

Chem had to go to work the following morning, so I ventured out on my own to explore the eastern part of the Vistula. I was rewarded by the sight of an utterly beautiful male Common Redstart flitting in a riverside bush, a Kingfisher perched on the opposite bank, a Common Sandpiper teetering along the muddy foreshore, tons more Fieldfares and Blackbirds, some of which were on nests, and three majestic White Storks soaring overhead. Black Redstarts are supposed to be common Kraków residents, with more than 100 breeding pairs, but I could only manage a solitary fleeting glance of a female.

I had arranged to meet Chem in town after lunch as he was going to break his walking habit and drive us nineteen miles out of town to visit Spytkowice Ponds, a large area of reed-fringed lakes in the Upper Vistula Valley. How could an urban birder turn down an opportunity for a rural birding fix? When we arrived I was not disappointed. I didn't know what to look at first: was it to seek out the Bittern that was booming in the reed beds next to me, the Great Egret that took flight nearby, the ten airborne Marsh Harriers or the numerous Whiskered Terns that were cavorting with Common Terns? My mind was officially blown when I listened to countless Sedge Warblers chirping away while a Great Reed Warbler knocked out its song and a Savi's Warbler reeled.

Kraków is a very interesting venue for urban birding, with exciting birds like Penduline Tit in Bagry, Grey-headed Woodpecker and Ural Owl in Puszcza Niepolomicka (a forest on the eastern border of Kraków) and good wintering waterfowl along the inner-city stretch of the River Vistula. It deserves more than a cursory glance and I'm definitely going to go back for a second look.

Lisbon, Portugal

During late October I was having the time of my life in the Tagus Estuary in the Lisbon region of Portugal. At over 37,000 acres, it was easy to see why its mosaic of habitats, which included farmland, marshes, muddy channels, lakes, salt pans and cork-oak woodland, was a birders' paradise. Almost every bush in this patchwork national park was heaving with amazing birds, and among the abundance of Crested Larks, Skylarks, Southern Grey Shrikes and House Sparrows were large numbers of 'Essex birds'. I saw more Essex birds here in Portugal than I've ever seen in Essex.

An Essex bird, I'll have you know, was my adolescent name for the Corn Bunting. In those days my trips to the Essex countryside were characterised by the sight and sounds of these drab-looking birds. I only seemed to see them easily in Essex. Since those heady days their fortunes in Britain have completely changed and their phenomenal decrease has been well documented. So I gladly took the opportunity to get reacquainted with this old friend.

Like many people, when I think of Portugal I immediately think of hot days in the Algarve, a round or two of golf (not that I play) and, of course, the rich birding to be had. So a late-autumnal trip to Lisbon was an interesting prospect, for although I had missed the bulk of the migration, perhaps I would be lucky enough to get the last proverbial squeeze of the tube.

I was a guest of the Turismo de Portugal and along with my guide, João Jara, they were very keen to show me the hidden Portugal that lay in between Lisbon and the Algarve to the south. They were also very apologetic for the potentially lousy weather forecasted and for the prospect of not seeing many birds. Apologies were wasted on me as I was already in heaven.

After a few days touring some really fascinating areas outside Lisbon, I finally ended up staying in the Expo area of the city quite close to the estuary and the stunning Vasco da Gama Bridge, which at nearly eleven miles from end to end is the longest bridge in Europe. According to João, Lisbon itself is not a hotbed of urban birding and he could only muster four sites worth visiting. Ordinarily, that kind of

statement would have made me even more determined to
seek out places to prove him wrong. But when you consider
the city's geographical position – right on top of one of
the most important feeding grounds for waders in Europe –
with such ornithological richness on your doorstep you
didn't need to go anywhere else.

However, we did visit Lisbon Gardens, a long, thin stretch
of reed bed and municipal parkland bordered by the long,
muddy foreshore of the estuary in the shadow of the Vasco
da Gama Bridge. We watched countless Black-headed Gulls
standing on the mud while joggers breezed past us along the
boardwalk. The gulls were at fairly close range and were
sharing their space with hundreds of Avocets, loads of
Redshanks, Curlews and Black-tailed Godwits, and several
hundred Greater Flamingos. We were in mid-count when
an old man in a cloth-cap approached us to speak of his dual
admiration for the flamingos as well as for the current
Sporting Lisbon squad. The reed beds, though rudimentary,
hold an array of warblers during the summer months,
including Great Reed Warbler, and in winter, Bluethroats
with Dartford Warblers in the scrubby areas. It's also a great
place for finding passage migrants.

Lisbon Gardens was just the appetiser for what was to
come when João and I crossed the bridge to enter the Tagus
Estuary birding areas. The sights that welcomed us will stay
with me for a long time. Scanning back across the estuary
towards Lisbon, I noticed a distant glistening thin pink line
along the shoreline of the city. A cursory glance through my
scope revealed that it wasn't a weird heat shimmer but
legions of Greater Flamingos. Some 5,000 choose to winter
in the Tagus along with at least 15,000 Avocets, thousands of
waders of several species, herons and egrets galore, plus
shedloads of Glossy Ibises and White Storks.

Every field seemed to host a surprise. I came across one
that contained at least forty exquisitely concealed Stone-
curlews, while another patch of land hid around seventy Little

Bustards that cautiously edged away from us. I looked at some bushes in one direction and counted about thirty Spanish Sparrows in a mixed sparrow flock, and in the opposite direction there were a few Rock Sparrows, more Dartford Warblers and singing Cetti's Warblers providing the background atmosphere.

Apparently, it gets pretty mad in the Tagus Estuary during the breeding season when hordes of migrants arrive, including the obligatory Mediterranean specialities like Bee-eaters, Great Spotted Cuckoos and Nightingales. João took me down a track on which during warm summer nights both European and Red-necked Nightjars churr side by side.

During the day that João and I spent at this wonderland we even found two Portuguese national rarities: a Marsh Sandpiper and a moulting summer-plumaged American Golden Plover. The opportunity for discovery is immense even 'out of season' and we never once met another birder.

As you can guess, if you like seeing Essex birds and want to go birding far from the madding crowd, then a visit to the Tagus Estuary is a must.

Merida and Cáceres, Spain

'Take me to the bridge' was a phrase made famous by the Godfather of Soul. I'm sure that when James Brown first grunted that line he wasn't thinking about birding, but it was the most appropriate phrase that sprung to mind on my first night in Merida, the capital city of the Extremadura region in Spain. I was in my hotel room on the phone to my guide, Martin Kelsey, who was briefing me on the following day's birding plans. He spoke of an ancient magical bridge literally ten min-utes' walk away from which the most amazing birds could be seen. It was no surprise that I couldn't sleep that night.

In the morning, after noticing a pair of nesting White Storks standing on an untidy pile of sticks plonked right on top of my hotel, we soon found ourselves on the bridge. Known simply as the Roman Bridge, it straddles the Guadiana River in the heart of the city and, at 790 metres, is the longest-surviving Roman bridge in the world. Aside from the river, which was a dangerous-looking torrent of brown water, the bridge overlooked a few reed beds that housed a male Little Bittern sitting right out in the open, an obliging Penduline Tit, many passage Chiffchaffs hunting for insects and noisy Cetti's Warblers.

It was early March, so it was strange to see migrant Swallows, and House and Sand Martins swooping around knowing that they had yet to make landfall in Britain. Local office workers walked past looking at us oddly as I whooped and gushed at the array of birds. Throngs of Cattle Egrets, with the occasional Little Egret in their midst, stood on the branches of clumps of trees that had become islands in a nearby egret roost due to the excessive floodwater brought on by the recent unprecedented rains.

A lone Night Heron flapped languidly overhead as a couple of Purple Swamphens stalked heavily around the remaining visible reeds with their enormous comical red beaks and ridiculously long toes. Martin explained that the Roman Bridge is the best place in the whole of Extremadura to see this oversized moorhen – and we're talking about a region the size of Wales.

Next was a forty-five-minute drive north to Cáceres, the capital of the Cáceres Province – the other of the two Extremaduran provinces. This city's centre is dominated by Moorish and medieval architecture, perfect for nesting Lesser Kestrels and White Storks. We stood on the elevated steps of a cathedral and were soon watching up to six Lesser Kestrels patrolling the tiled rooftops with an elegance that blew me away. They twisted, turned and swooped with breathtaking effortlessness. Their colonial habits, altogether brighter plumage with less streaking and a slight central tail projection separate them from the Kestrels that we know and love.

When these adorable birds were out of view we were entertained by several Spotless Starlings whistling their simple, watered down Common Starling-like songs, flocking Jackdaws and feral pigeons swirling around the rooftops and, of course, the graceful gliding and awkward landings of several of Cáceres' 250 pairs of White Storks. In the sky above the occasional migrating Red Kite passed over often in the slipstream of its more numerous Black cousin. Now that was what I call urban birding.

Our final urban destination was the small town of Trujillo, another forty-five minutes on from Cáceres. Here, we visited a bullring in a fairly rundown area at the edge of town. On the surrounding derelict land Crested Larks, White Wagtails, House Sparrows and Serins hopped among the weeds, while overhead were many Swallows and House Martins with the occasional Red-rumped Swallow.

But it was the bullring that we had come to see because up to thirty pairs of Lesser Kestrels nest on the doughnut-shaped stadium roof. We were treated to spectacular views of the birds in courtship mode and watched the females entering the crevices of tiles or the specially supplied ceramic nesting chambers while their mates seemingly stood guard. The great thing about all three of the cities I visited was that they hold the unusual distinction of being among the very few urban areas in Europe designated by the EU as Special Protection Areas for birds. Extremadura has about fifteen per cent of the entire European Lesser Kestrel population, and two of the largest concentrations of breeding pairs are in Cáceres (about 250 pairs) and Trujillo (about 145 pairs).

Of course, I could not come to Extremadura without sampling the birding out in the wilds, and under Martin's excellent guidance I managed to catch up with a panoply of goodies: Great Spotted Cuckoo, Thekla and Calandra Larks by the bucket load, Blue Rock Thrush, migrating Common Cranes, Black Storks, Spanish Imperial, Golden and Short-toed Eagles, and masses of Griffon Vultures with lesser numbers of Black and Egyptian Vultures. The list was endless.

On my final morning I returned to the Roman Bridge where it all began. Unfortunately, the water levels had risen further to submerge the scant vegetation that was visible only forty-eight hours earlier, plus the skies were grey and threatening yet more rain. I took solace in watching a Crag Martin that was aerobatically patrolling the length of the bridge. I watched its every twist and change in direction and thought, if James Brown were a birder what line would he have written to describe this?

Paris, France

My quest to deliberately target the cities that are deemed to be birdless concrete wastelands inevitably included Pais. Classically, the French capital is a place where you are more likely to experience a romantic interlude, pose grinning in the shadow of the Eiffel Tower or get bored senseless at some dull business meeting rather than find birds. Or so it seems. I must admit that my experiences of Paris up until now have been purely for business and sadly not romantic or bird rich – my time in the city has been incredibly short like my resultant bird list.

So, at 5.25 a.m. on a slushy February morning fresh after the worst snow event in southern England for eighteen years, I boarded the Eurostar at St Pancras to head directly into a weather system drifting across central France that promised gallons of rain. Questions had to be asked about my sense of timing. I had nowhere to stay, but at least I had a guide for my first day in the shape of Maxime Zucca, a young Parisian birder who answered my very last-minute Internet plea for help. Yes, I really do believe in forward planning.

Once in Paris, I braved the rush-hour metro and within minutes of meeting Max on a grey street corner in the south of the city, he had invited me into his nearby apartment to meet his rather attractive girlfriend, who was also a birder. They made plans for me while the kettle boiled and toast with plum jam was devoured.

Soon, Max and I were out on the city streets and like a true urban birder he looked at everything that possessed wings, stopping to watch over some odd spots like his local section of the Chemin de fer de Petite Ceinture – a largely disused rail track that circumnavigates Paris. As he eagerly peered down onto the railway sidings, he explained that the embankment vegetation held breeding Spotted Flycatchers and some of the common warbler species. He fondly recalled the day when looking onto the tracks he startled a Woodcock, and gushed about the Ring Ouzel that flew over his head while he was skywatching from the top of a nearby bus station.

I was fascinated to learn that species like Carrion Crow and Sparrowhawk were relatively new arrivals to the Parisian avian scene and that the local birders were eagerly waiting for the first Peregrines to come and take up residence. As we walked through the grounds of the Cité Universitaire and Parc Montsouris, Max proudly pointed out the nest hole of the city's only breeding pair of Rose-ringed Parakeets. I shot him a withering glare as the first drops of rain began to fall.

Max then took me under a tree and started 'phishing'. We had a Dr Dolittle moment shortly after. Out of nowhere, the tree literally filled up with birds. Aside from Blue and Great Tits, there were a couple of Crested Tits, Nuthatches, a Lesser

Spotted Woodpecker and a ridiculously close-up Short-toed Treecreeper. I looked at Max in wonderment. We took in a few other parks including Jardin de Luxembourg and Jardin des Plantes, where we saw more Short-toed Treecreepers and tits, plus the additional excitement of two Hawfinches (a city scarcity) flying over the Jardin de Luxembourg in the pouring rain. Oh, did I mention the rain? By the time that we got to the nearby Seine, I was sodden, and carrying a by-now-wet rucksack that was beginning to concern me as it contained my laptop. Before calling it a day to find a hotel, we gave the river a scan resulting in some Grey Wagtails, a Yellow-legged Gull and, most surprisingly, a Kingfisher – all right next to Notre Dame.

The next morning after checking out of the dilapidated hotel that I had found the night before, I headed south-east to the Bois de Vincennes, a large area of woodland traversed by many paths and man-made lakes. Both this wood and the Bois de Bologne in the south-west of the city are great spots for woodpeckers. With the wind and rain whipping up I decided to cut my losses and head to Paris's secret jewel – the Parc des Beaumonts.

This small park is set on an incline and completely surrounded by blocks of flats and a cemetery. Initially, I was very disappointed because it seemed like a boring landscaped inner-city open space. Was my urban birding radar faulty? Was this site a turkey? My disappointment turned to glee, though, when I reached the brow of the hill, as the rest of the park was a mixture of woodland and scrub managed for nature, with a tiny reed-filled wetland that sported a solitary Moorhen.

Parc des Beaumonts apparently comes to life during passage times, as species like Black Stork, Middle Spotted Woodpecker and Marsh Warbler have previously shown. There's clearly a lot of promise here – if you can ignore the potential hoards of people. My Parisian adventure ended back in the city centre at Cimetière du Père-Lachaise, the

final resting place for such cultural luminaries as Chopin,
Jim Morrison and Oscar Wilde. Despite the rain, I managed
to discover a pair of Marsh Tits, a local rarity.

The birding scene in Paris is still quite small and there is
such great potential to find really good birds. Next time you
are in this swish city, bring your bins and check out its
hidden treasures.

Prague, Czech Republic

The prospect of going birding within the confines of a city is
an interesting one. It is a complete dichotomy: watching for
wildlife in the heart of an urban centre. Surprisingly, some
cities are as under-recorded for their birds as the remote
corners of the Earth. Although most of the hitherto-unknown
species lurking in our cities are generally very familiar and
are often by no means globally threatened, they are still
minor miracles that should be celebrated and protected.

One such city where the birdlife is not generally well
known is Prague. The architecture doesn't get much more
stunning than in this historic capital; it's a great sightseers'
city. Although better known these days as one of the chief
venues for drunken stag-party weekends, I had the pleasure
of visiting Praha (as it is known locally) for a couple of days
during August on a non-birding trip to see some Czech
friends.

Given that my time was short, I was unable to give this
city a real going over. I didn't even have a chance to check a
cemetery, one of my favourite urban haunts. Prague has great
potential as many of its parks support great birds, including
Marsh Warblers that supposedly breed around the city zoo.

But it was the prospect of those eastern woodpeckers that excited me most.

I stayed in a hotel in the Old City, a short walk away from the Vltava River that dissects the city. From my hotel window I had great views of Malá Strana, basically a posh district of the town across the river dominated by a wooded hillside park called Petřín. After meeting up with my friends for breakfast I headed over on foot to Malá Strana. On the way I crossed over the famous Charles Bridge, pausing to watch a few Black-headed Gulls patrolling the river over the heads of a large herd of semi-domesticated Mute Swans. The river attracts an array of wintering waterfowl and, according to the Czech Society for Ornithology, it is one of the country's most important wintering areas for waterbirds such as Pochards, Tufted Ducks and Coots. Recommended, though, is the stretch of the river in Prague-Trója, which is in the vicinity of the zoo. Apart from the commoner waterfowl you could also expect Teal, Goosander, Goldeneye and scarcer visitors.

When I arrived at Petřín it reminded me of a large, heavily wooded version of London's Holland Park, with the hill itself crowned by what looked like a mini-Eiffel Tower. There were plenty of strolling families and tourists traversing the many paths, and a light railway/tram provided an alternative mode of transport through the park. Despite the apparent disturbance I was surprised by the abundance of birds. A couple of Turtle Doves flew over as I was scrutinising the drifting Swifts. Green and Great Spotted Woodpeckers entertained me as I searched in vain for the woodpecker species that help to make this part of Europe so special.

I was stunned by the close approach that some of the birds allowed me. I practically walked up to a female Common Redstart as she quivered her tail on a branch over my head. Short-toed Treecreepers were common as were Marsh Tits. I saw what I thought were female-type Pied

Flycatchers but I was aware that the very similar Collared Flycatcher also frequented the area. I wasn't able to discern the latter's paler rump band and bigger white primary patches, so I cautiously opted for the commoner species. The Hawfinches were ridiculously approachable. I watched a female and her offspring out in the open less than fifteen feet away from me. It was a weird feeling watching a bird that we know at home as extremely shy and retiring, going about its business with no regard for me in the middle of a city.

Perhaps the most surprising bird I saw was a juvenile Honey-buzzard that I first noticed quietly sitting in a tree trying to hide from me. Initially, all I could see was its brown head and neck. When I raised my bins, I noticed its dark eyes and yellow cere. It obviously decided that I had seen enough and lazily flew off, landing clumsily in foliage some distance from me. Its size was remarkable and its jizz immediately struck me as being kite-like.

The following morning I headed out to Divoká Šárka, an area of forest country park on the outskirts, some twenty minutes' taxi ride from the city centre and not a million miles away from the main airport. I only had a few hours, so my goal was to try to clean up on the woodpeckers that I so wanted to see. On arrival I was greeted by the sight of a pair of Common Buzzards soaring on thermals. Walking through the forest I had great views of several singing Wood Warblers and roving flocks of the Central European race of Long-tailed Tits. They sported frostier-looking heads than the British race. But the woodpeckers that I craved were nowhere to be seen.

Dejected, I headed back to where my taxi was waiting. On the way I came to a forest clearing, so I decided to give my woodpecker hunt one final effort. Within moments I heard the familiar 'yaffle' of a Green Woodpecker. I didn't come all the way to Prague to see a flipping Green Woodpecker, I snorted under my breath. As I was about to give up, I noticed a dark, crow-sized bird flying towards me in a flight pattern that was a cross between that of a woodpecker and a Jay. It was a Black Woodpecker and it flew right over my head.

Reykjavík, Iceland

There's a certain amount of swimming upstream involved if you want to be an Icelandic birder. There are only around 370 species on the country's list and almost all of the regulars either occur in its capital city, Reykjavík, or have shown up as scarce visitors. Icelandic ornithologists don't see themselves as birders at all. They call themselves finders. Apart from having an ingrained propensity for surveying, they also actively seek unusual visitors to their shores. I loved the term finder because that's the kind of birding I'm into – finding new locations and new birds. If you lived in Iceland, it would be possible to see all the regular birds fairly quickly. Once you've seen everything you will then moult into a finder.

Don't get me wrong, despite the relatively short list the Icelandic birdlife is nonetheless incredible and in Reykjavík it is no different. Where else can you be in an urban setting and watch Red-necked Phalaropes going about their business and hear the hauntingly chilling wailing of Great Northern Divers – sampled in many a Hammer horror film and by cool house-music producers?

My Reykjavík guide was the James Corden look-alike Hrafn Svavarsson. His christian name was as hard to pronounce as it looks. When uttered by anyone with a hint of an Icelandic accent it sounded like a cross between the deep 'krrop' call of a Raven and someone clearing their throat. Indeed, Hrafn is Old Norse for Raven, so naturally he had to be rechristened Raven. Even though it was mid-June, the height of summer and the land of the midnight sun, I still found it surprisingly nippy as I stood on the shoreline of Elliöavatn at 1 a.m. in broad daylight. Hrafn was triumphantly showing me a pair of nesting Great Northern Divers as we listened to nearby singing Redwings. It was a pretty big lake with plenty of Red-necked Phalaropes, Scaup, Tufted Ducks and Teal to keep our attention, while on a nearby rocky mount stood a male summer-plumaged Ptarmigan, a species that I saw a lot of over the ensuing days. Apparently, the lake is one of the best places to try for vagrant Grey Herons and the odd Swallow, the sorts of birds that we take for granted here in Britain. I was constantly looking up expecting to see wafting Swifts or wheeling hirundines, but instead I saw empty skies bar ever-present drumming Common Snipe.

Bakkatjorn, a small lake on the Seltjarnarnes Peninsula, west of the city centre, looked like a particularly interesting spot to me. Hrafn confirmed my hunch when he said that this was the rarity hotspot of the city, as finders like to gather here in the autumn. Despite being urban, there were loads of nesting Arctic Terns everywhere, plus on the far shore of the pond stood several score of Glaucous Gulls among the other more familiar large larids. A golf course adjoins the site to the west and holds good numbers of Brent Geese in the winter, whereas the lake attracts Red-breasted Mergansers and many yodelling Long-tailed Ducks. But as I mentioned earlier, it's the potential for vagrants from both sides of the Atlantic that draws people here. The list is as long as it is varied and includes such beauties as Bonaparte's Gull, American Wigeon and, from closer to home, Common Tern and Long-tailed Skua.

For those who visit Reykjavík for business or drunken stag/hen weekends and have little time for any birding, Tjörnin, right in the middle of the city, is for you. Despite having small areas of more manicured parkland, it's a lake surrounded by relatively untouched marshland, a snapshot of the type of habitat found in abundance elsewhere in the country. Arctic Terns are ubiquitous, Snipe breed here, Water Rail used to until the Mink released in the Sixties by unscrupulous individuals extirpated them, and there is the usual array of Reykjavík ducks with a healthy spattering of Greylags. Winter is especially nice here, with Long-tailed Ducks sometimes consorting with the occasional Pintail. Rarities show up here, too, and there's the odd record of a Greenland Gyrfalcon and Ring-necked Ducks sometimes make an appearance.

As mentioned earlier, burial places are among my favourite inner-city habitats. Fossvogskirkjugardur Cemetary, situated south of the city centre near the coastline, didn't disappoint as it harboured a lot of the passerines that may not necessarily be easy to see at the other sites that I have discussed. Iceland's hoary-looking Redpoll is common here. They were very white looking, although not white enough to be vagrant Arctic Redpolls. Redwings were the default passerine, while Hrafn got quite excited when we heard a Blackbird sing. I almost overlooked it, being so used to them in Europe, but here they are very scarce and treasured residents. It goes without saying that finders have done their work here and have picked up errant migrant Chiffchaffs, Willow Warblers, Bramblings and Greenfinches over the years. They are big birds for Iceland.

Reykjavík may currently be expensive, but on the plus side there are no mosquitoes, the people are very warm and have a dry sense of humour, and if you visit you will have a great opportunity to get close to large numbers of the species we only usually get to see sparingly in Britain. My overriding Reykjavík memory was watching the incredibly numerous Arctic Terns that were practically everywhere. They were so captivating. I don't think that I will have a problem recognising one ever again.

Valencia, Spain

Have you ever picked up a national newspaper on a cold winter's day and come across one of those cheap, no thrills airline adverts purporting to offer flights to a ton of European destinations for just a few pounds? It happened to me recently. I looked down the list and Valencia got the vote. I didn't know much about the city other than I liked its football team, but I knew it would be a good bet for the expected Mediterranean bird species. So I booked the flight.

Six days and £180 later I was stepping off the plane in Valencia to begin a four-day vacation in Spain's third largest city. Situated on the east coast about 185 miles south of Barcelona, Valencia has a semi-arid climate and, indeed, the hot afternoon sun was blazing down, which was a far cry from the chilliness I had left behind in London. Not normally one for planning, I had decided to make the most of my city break by heading directly to Albufera de València, seven miles south of the city. It was once one of the largest expanses of coastal marsh and wetland in the country, but thanks to years of reclamation, urban development, pollution and tourism it has taken a battering. The area was declared a national park in 1986 and when I discovered that its present size of some 44,400 acres was roughly a tenth of its former glory, I had to admit that I felt a bit glum.

After getting a hire car I drove to a B&B in Benicull, a small village near the marshes that was run by a classic expat couple who quite frankly looked like a pair of retired bank robbers. The wife, whom I only saw the once (when she opened the door to me), was in her sixties, although she dressed at least twenty years younger and had heavily bronzed, leathery skin. Her husband looked as though he had stepped straight out of an *EastEnders* set and had the classic 'Great Train Robber' look. I quickly realised that he was a tad racist, a fact that he barely concealed when he first saw me. This manifested itself in the ensuing mornings when he would serve me a hard-boiled egg and toast while he privately knocked himself out a full English in the kitchen. Regardless, I made myself comfortable and on my arrival afternoon decided to go for a stroll around the plentiful orange groves that enveloped the area. I quickly notched up many Nightingales – some were singing away right out in the open. Serins abounded and in the air were Common Buzzards, the occasional Kestrel and many Grey Herons, and Cattle and Little Egrets. Yes, this seemed like a lovely local patch.

The following couple of days were spent exploring the expanse of Albufera de València. I found that it was a mixture of rice fields, grazing meadows, sailing centres and a nature reserve that included a fair portion of the area's standing water. In the early mornings I gazed skywards to watch the troops of herons and egrets incrementally leaving their roosts. Aside from the species that I had already discovered back at my adopted local patch, I also witnessed Night and Purple Herons flying to destinations unknown, along with the occasional Squacco Heron. On land there was a multitude of birds to look at, with House Sparrows being the default passerine throughout the region. Tree Sparrows were also around, though I may have overlooked the vast majority. On one occasion I did glimpse a male sparrow that appeared to have faint dark streaks on its flanks, but I didn't see it long enough to commit to calling it a Spanish Sparrow.

It was early spring so migration was pretty evident. I remember seeing a Black Kite casually drifting in from off the sea and watching it for at least ten minutes as it meandered through. I did see quite a few Pallid Swifts with their common cousins, and could pick out their frenetic flight and dark 'sharper' shapes. All the Hoopoes I saw were invariably flying away from me, while Turtle Doves were reassuringly common.

But it was the waterbirds that made the biggest impression on me. There were plenty of noisy Black-winged Stilts in the nature reserve, while swooping overhead were at least

twenty Collared Pratincoles hawking for insects among the
Swallows, House Martins and Swifts. One afternoon I was
lucky enough to find a migrant drake Garganey and a Little
Ringed Plover on the shoreline. From the hide I could see
secreted among the hundreds of nesting Black-headed Gulls
several pairs of Mediterranean Gulls, a couple of Audouin's
Gulls and a lone first-summer Little Gull that spent most of
its time hiding behind the larger Black-headed congeners.
Standing around the colony like sentinels were several
Yellow-legged Gulls waiting for an opportunity to have a
late lunch.

My favourite moment, however, was discovering a
displaying pair of Slender-billed Gulls on the water right
outside the hide. It was the first time I had ever really studied
this long-necked species close up – the size difference was
very clear, with the male being ten per cent bigger than the
female.

Valencia is a must for the birding but try to avoid any
retired convicts.

The Rest of the World

Addis Ababa, Ethiopia

Have you ever indulged in nude birding? I nearly found myself doing just that in early May on my first morning in Addis Ababa, the capital of Ethiopia. I had arrived in the city under the cover of darkness during the previous night, an excited participant on a joint BirdLife International/ RSPB mission to survey four of the country's most enigmatic and rarest endemics: the handsome Ethiopian Bush-crow, the glamorously titled Prince Ruspoli's Turaco, the mysterious White-tailed Swallow and, rarest of all, the little-known Liben Lark. Before I could be let loose into the Ethiopian hinterland the plan was for me to serve my Horn of Africa birding apprenticeship by spending some time getting used to the plentiful avifauna of this intriguing country in Addis.

I remember waking up with bright hot sunlight beaming into my Spartan ground-floor hostel room and hearing big cockroaches scuttling in the bathroom shower tray. More worryingly, I also heard the sounds of members of my party already birding outside my door. Without thinking, I stumbled out of bed and raced outside into the hostel garden. I was immediately subjected to a sensory overload. It felt as though I was watching a virtual slideshow; there were birds everywhere. A Brown-rumped Seedeater morphed into a Mountain Thrush as Nyanza Swifts swilled overhead in the same airspace as traversing Pied Crows and Dusky Turtle Doves. It really was a classic case of not knowing what to look at first and having blurry sleepy vision didn't help matters either. I did manage to glimpse what was almost certainly a migrant Lesser Whitethroat deep in the foliage of a nearby small tree – a breath of familiarity in what I could only describe as an avian wonderland. It was then that I realised that I had rushed out in just a pair of underpants and nothing else, so I beat a hasty retreat to get dressed and, crucially, grab my binoculars.

Addis Ababa is an excellent urban birding hotspot and is a place where you could discover almost twenty of the country's endemic species, with birds like the Brown-rumped Seedeater that I previously saw being more easily found here than almost anywhere else in the country. After a move to a far more comfortable hotel I spent the next few days roaming the city and its environs looking for urban-birding sites under the watchful eyes of multitudes of circling Yellow-billed Kites, and Hooded and White-backed Vultures. I visited the inner-city gardens of the Ghion Hotel where many Ethiopian feathered beauties revealed themselves to me. Some were totally unrecognisable, but the Rüppell's Robin-chats that flashed past me into dense cover like large bright Common Redstarts, and the foraging Tacazze Sunbirds, the males resplendent with their long tails and ridiculously iridescent plumage, were a little easier to identify.

The Entoto Hills, on the northern outskirts of the city and a short cab ride from my hotel, were an oasis for a range of more montane birds. Moorland Chats were abundant and easily approached. These cute little brown birds with short black-and-white tails seemed like the crazy love children of a Robin, a Northern Wheatear and a Stonechat. Perhaps the star bird up in the hills was a fine, tawny-looking Wahlberg's Eagle that briefly drifted over. I discovered other great spots back in the city like Yeka Park and Bihere Tsige Public Park. Both were maintained botanical gardens that charged small entry fees (and an additional amount if you were wielding a camera of any description), but it was worth it because they were rammed with exotics like Blue-breasted Bee-eaters, Nubian Woodpeckers and White-backed Black Tits.

However, perhaps the most exciting place I found was also the most unlikely area for birds. I was birding from the hotel rooftop, watching the drifting Hooded Vultures after cheekily asking reception if I could venture up there. I had expected a resounding no but I wasn't in the UK, so health and safety was not such a big issue in this city. Eventually, a maid took me into a top-floor bedroom and through the window onto the rickety-looking corrugated iron roof. Mindfully, I had grabbed a pillow from the bed on the way as I knew that there would not be any comfortable seats on the roof. It was up there while seated on my pillow that I noticed a Blue-breasted Bee-eater land on a telegraph wire next to an interesting piece of rough ground. I immediately had to go and investigate and fell in love with what I saw.

It was a fifteen-acre area of waste ground that lay behind my hotel. Effectively a building site, it was covered in rubble, thistles and other scrubby invasive vegetation. I also noticed that it was attended by a host of locals who saw my adopted local patch, which I visited twice daily, as one big alfresco public latrine. Although the stench was often suffocating, the local flies saw it as manna from heaven, and I managed to find some amazing birds. Every day heralded a surprise, including the discovery of many migrants destined for Europe. I found several very pale eastern-race Common Whitethroats, Sedge Warblers and an exceptionally grey-and-white Willow Warbler all hopping on and around the human waste. No doubt all those tasty flies attracted them. There was also a pair of resident Common Fiscals and Groundscraper Thrushes that patrolled the area every day along with other local species. An absolutely stunning male Red-backed Shrike frequented the scrub for a couple of days and a gorgeous male Isabelline Shrike also put in an appearance one afternoon. It was hard to watch that beautiful bird with the unsavoury backdrop of a guy squatting in full view with his trousers around his ankles.

I saw more than sixty species on my smelly patch and even watched an endemic Wattled Ibis swallowing lumps of faeces with relish – double yuk! But it goes to show that no matter where you may find yourself in the world and no matter how unsuitable your patch may seem, if you watch a place regularly you should be prepared to be pleasantly surprised.

Bangkok, Thailand

I woke up one morning just before Christmas and realised that I was totally knackered. Chasing around the place urban birding and giving talks had all taken its toll. So, after months of canvassing by my girlfriend I finally allowed myself to be talked into taking a holiday. She invited me on her annual Christmas pilgrimage to visit her sister's family in Bangkok. Let me explain, my girlfriend's not a birder, though in common with many non-birder partners, she tolerates my obsession. She has about two hours a week of interest in her before she hits a wall; often without warning. So, as a compromise, I promised to put down my binoculars and not disappear for the two weeks chasing shadows in a jungle, but instead to pick up a tourist map. I had a feeling that it was going to be a tough call for me.

Bangkok is a big city, with a population comparable to London. I found it to be a fascinating place: bustling, vibey, hot and there were birds aplenty. I remember discreetly clocking up Red-rumped Swallows whirling in the sky alongside their commoner Barn cousins, while jet-lagged and flopped out on a deckchair in my girlfriend's sister's garden on the outskirts of the city. Above my head in the trees was a foraging Yellow-browed Warbler, calling as it gleaned minute insects from the foliage. There was also a Pied Fantail flitting around with tail fanned, a couple of frolicking Oriental Magpie-robins and, reminding me of home, a Common Kingfisher was perched on a pole in the lake that bordered the garden. All was going to plan. I was holding back on the birding and she was happy. Result.

Temptation didn't take long to find me. The next day Dave Gandy, an old friend and an expat resident of Bangkok, contacted me. He was supposed to be away for Christmas but his plans had suddenly changed and now he was tempting me out on a pre-work scout of Suan Rot Fai, also known as Wachira Benchathat Park or Railway Park, his local patch in the north-west of the city. To assuage any potential ire, I brought my girlfriend with me. We all spent a couple of fantastic hours exploring his urban paradise, which was still recuperating from the recent floods that had hit the city in the previous weeks. It was a well-trodden park with over 140 acres of lightly wooded areas latticed with creeks. The place was rammed with birds ranging from the ubiquitous Tree Sparrows to Black Drongos and a handsome Black-winged Cuckoo-shrike. Dave was pointing out the diagnostic features of a Taiga Flycatcher perched low in a tree when a gorgeous Indian Roller, resplendent in royal blue, flopped out of the same tree. Later, we even twitched a wayward Chinese Blue Flycatcher that had been in the area for the past few weeks. Dave has recorded 114 species in the two years he has birded the park, including some incredible local rarities. I was well impressed with this unlikely urban gem of a patch.

All the 'pretty birds' at Suan Rot Fai captivated my girlfriend. I was in her good books and all was well. A few days later we journeyed 124 miles south to spend some time at Hua Hin, a coastal resort, where my girlfriend summarily fell ill with tonsillitis. Our resort was conveniently situated an hour south of Pak Thale, the world-famous salt-pan wader hotspot that is also the winter hang-out for a tiny handful of one of the world's rarest and most enigmatic birds – the Spoon-billed Sandpiper. So for two mornings while my beloved lay stricken in bed, I took a taxi to this wader paradise and revelled at the sight of hundreds of Marsh Sandpipers, Spotted Redshanks, Red-necked Stints and good numbers of other waders like Pacific Golden Plovers and Terek Sandpipers. Despite finding a couple of site scarcities, a Red-necked Phalarope and an Avocet, I could not find a 'Spoonie' largely because I didn't have a telescope. Leaving Thailand without seeing one was not an option, so I needed to call in the big guns.

Enter Phil Round, Thailand's top lister, rediscoverer of the once-thought-extinct Gurney's Pitta and all-round hot birder. I convinced my girlfriend that we should leave Hua Hin that morning, a day ahead of schedule, and on the way back to Bangkok meet up with Phil at Pak Thale, twitch the Spoonies and arrive back in the capital for lunch. Well, twelve hours, five Spoonies, a Black-faced Spoonbill and countless other ticks later, we finally arrived in traffic-stacked Bangkok. It was nightfall and I was in the biggest doghouse known to man.

For the remainder of my holiday I had to deny any interest in birds in order to preserve the peace. I did manage to sneak a quick visit to Bang Poo just outside Bangkok. It had a seaside promenade not dissimilar to Brighton's where the locals gathered to feed thousands of braying Brown-headed Gulls not with bread but with scraps of noodles and prawns. During the short time I was allowed out, I found six Black-headed Gulls in their midst and an overflying Osprey but no Pallas's Gull – an impressive gull that is occasionally found there.

I thoroughly enjoyed my Bangkok holiday and could have written several pages about the birding to be had. But first, I'd better start working on my girlfriend to convert her into a birder.

Cape May, New Jersey, USA

I crawled through the baggage hall at Philadelphia International Airport feeling lower than dolphin poop. I had just endured a hellish five-hour flight besieged by stomach cramps and other related ailments due to the horrible virus I had caught in Tucson, my previous destination. To add insult to injury, the cheap hire car I thought I had booked to drive the ninety minutes to Cape May, New Jersey, wasn't waiting for me. I was due to stay with birder and photographer Richard Crossley and his family for a few days and it was getting late. They were expecting me. It was night, freezing cold and snow liberally coated every surface apart from the roads themselves. Knackered and sick, I bit the bullet and got ripped off by an airport hire-car company to organise last-minute transport.

After being stopped by the police for making an illegal turn – this time pleading my Englishness worked – I arrived at Richard's house at 9 p.m. Richard is an ex-pat Yorkshireman who settled in Cape May a few years ago after falling in love with the area and the birding. He is most famously known for his recent opinion-dividing and mould-breaking photographic guide to the birds of eastern North America. For years I was drawn to Cape May, one of the USA's foremost birding hotspots, after hearing about the legendary falls of migrants of many species. Birding is also pretty damn good throughout the year, as I found out despite it being February, and freezing cold, and feeling that my time on this planet was swiftly drawing to a close. Watching Richard's garden feeders for five minutes proved that. I saw my first Carolina Chickadees as they fed alongside fellow Tufted Tits (or Titmice if you are American), stunning Northern Cardinals and hulking ground-loving Fox Sparrows.

Cape May is a curious place architecturally. The general theme seemed to be based on a Victorian seaside town; indeed, it is the US's oldest seaside resort. Jutting out into Delaware Bay, it is a fantastic magnet for neotropical migrants. I always imagined that the birding was restricted around the Bird Observatory in the northern end of the Cape. However, I was totally wrong, as Cape May is a huge area with a multitude of places to go birding, ranging from nature reserves to the very streets themselves. It was absolutely my kind of birding.

For instance, the visit that Richard and I made to Wildwood Crest Coastguard Base was an interesting one. Previously out of bounds, it is now a nature sanctuary where people are free to roam its coastal scrub and woodland, and to explore the beach. It had buckets of birding potential, and Richard had high hopes of finding something interesting there one day, although the large roving flocks of American Robins were good enough for me.

We found some great birds in the local vicinity, including waterfowl like Hooded Merganser, Bufflehead and properly wild Canada Geese. In the trees, aside from the abundant roving American Robins, we also came across urban feeding groups of delightful Cedar Waxwings and terrestrial Red-winged Blackbirds. In the coastal evergreens were flocks of Crossbills that allowed very close approach, plus we even happened across a couple of Snow Buntings and a Tundra Swan – a local scarcity. A lot of the habitat is taken up by a coastal marsh that in places is up to four miles wide and stretches some 100 miles to just south of New York in the north. This marsh provides nesting and resting places for birds as diverse as Clapper Rails and Seaside Sparrows.

Spring in Cape May is heralded by the return of the first Laughing Gulls that arrive as early as late February. By the end of March things start to pick up when the first of the wood warblers touch down: Louisiana Waterthrush and Yellow-throated Warbler are the classic early migrants, which actually stop to breed. By the third week of May everything is peaking, including masses of Horseshoe Crabs that congregate on the beaches to lay eggs. These attract a Knot, Sanderling, Turnstone and Semipalmated Sandpiper feeding frenzy. Many species of wood warbler pass through in their resplendent breeding plumages, alongside handsome Rose-breasted Grosbeaks and Indigo Buntings. It is a riot of colour and birdsong. Summer is a nice time to visit but it is also pretty popular with human visitors flocking to the beach and potentially invading the breeding sites of the endangered Piping Plovers and Least Terns.

But it is in autumn that you really need to be birding the area. It starts in late June when the climax of passage passerines kicks off during September with up to thirty species of wood warbler in a week. October is the month for the most raptors, and during the end of that month the major sparrow passage occurs, with flocks that can sometimes number in the hundreds of thousands. I heard an incredible story concerning a particularly heavy passage of autumn American Robins. Over two million birds passed through during one day and observers standing in the street witnessed the astounding sight of a continual stream of birds flying down the road at waist height and between their legs.

It was after I'd waved goodbye and was dragging myself to the airport that I heard about the most defining Cape May moment. Imagine this: a birder driving through the town early one morning watching a guy walking down the street inadvertently flushing hundreds of migrants from front-garden bushes. The birds scattered in wavering multicoloured flocks as if they were autumn leaves hurled into the wind to waft like confetti. What an image.

Chiang Mai, Thailand

I love Thailand, not least for its birding and wildlife. I also think that the Thai people are smiley, friendly folk possessing humility and a wicked sense of humour. I certainly thought about the latter attribute when I read some of the name tags worn by the staff in my hotel in Chiang Mai. Dhum, Prik and Porn had me chuckling. Their parents must have been pretty pie-eyed when those kids were named.

I had visited Bangkok the previous Christmas with a great degree of success when it came to urban birding. Plus as a bonus, I managed to cop sight of five of one of the world's most endangered birds – the Spoon-billed Sandpiper. Well, this time I thought that I would try a different Thai flavour and head further north to near the Myanmar border and visit Chiang Mai.

The city had a very different character from Bangkok. Set at the foothills of the Thanon Thong Chai mountain range, it is a smaller and more arty city stacked with temples. To the north is Doi Inthanon, which at 8,415 feet is Thailand's highest peak. Thai holidaymakers make an event of getting to the peak for dawn to experience 'being cold' and seeing frost.

The hot birding spot there is Chiang Dao Wildlife Sanctuary, apparently one of the best places in the country for the sheer variety of birds. However, the two species birders really hope to see there are Giant Nuthatch and Mrs Hume's Pheasant. On an early-morning ascent on the back of a pick-up truck the 'Lucky Lindo' syndrome struck again. The first bird we saw was a Large-tailed Nightjar flushed in the half-light from the road before us. The second species compromised two separate male Mrs Hume's Pheasants, despite their reputation for being difficult. One bird momentarily sprinted alongside the vehicle and the other ran onto the road in front of us to issue a quick stare in our direction before bolting. Looking like a grey-tailed, washed-out version of our own familiar pheasant, this bird is a real rarity with apparently only a third of the visits to this reserve resulting in sightings. Before we had even hit the reserve proper Giant Nuthatch was in the bag. It certainly was a whopper. Imagine a Nuthatch the size of a Song Thrush and there's your Giant Nuthatch right there.

After getting my listing fix I spent the remainder of my five days in the city proper getting to grips with the urban birding. Indeed, the first Chiang Mai species that I recognised was a gorgeous Purple Sunbird that conveniently sat at close range in a tree next to my cab as I emerged from the airport. I'm now glad that I spent some time looking at it because it was the only one that I saw. The same could not be said of the abundant Yellow-browed Warblers, whose incessant 'serri-serri' call could be heard from almost every tree regardless of location. Looking at the hotel map I noted that there appeared to be little or no green spaces nearby, so I took a cab one morning to Huay Tung Tao on the outskirts of the city. It's a lake surrounded by a well-traversed woodland, both of which are under the jurisdiction of the military. The site provided me with a number of new birds, including a roving flock of Olive-backed Pipits – a bird that I had always dipped out on in the UK. I had to vacate the area by midmorning because it had become overrun with boy scouts, girl guides and, of course, the army.

Back at base, I decided to generally explore my local area in the vain hope of chancing upon a potential local patch. At first my urban exploration wasn't up to much. The nearby River Ping ponged. No life there, bar a few hawking Barn and Wire-tailed Swallows. The exploration continued and included accidently walking into a knocking shop thinking that it was a shopping centres – long story.

Eventually, I found an area of littered scrubby marsh interspersed with a few mature trees and lots of barking feral dogs. A water lily-chocked stream ran through it and the whole site was surrounded by habitation, but especially on the eastern side where a large, ugly housing block stood. Perfect, and just ten-minutes' walk from the hotel. Over three mornings I recorded nearly forty species ranging from a resident Taiga Flycatcher (a recent split from the Red-breasted Flycatcher) to elusive Moorhens that crept around the scrub with White-breasted Waterhens. Among my favourite sightings were the delightful groups of pastel-coloured Chestnut-tailed Starlings. They flew around in small flocks, descending on fruiting trees like a pack of Waxwings, frantically gobbling up berries.

But the star of my urban patch had to be the Asian Barred Owlet that I found sitting on an exposed branch on a tall tree. Don't let the name owlet confuse you. This bird is a brute. The crescentic barring on its face, which narrowed in to frame its fearsome yellow eyes, did little to add to its 'ah' factor. Once the bird knew that I had my binoculars trained on him, he dropped out of the tree and powered into ano-ther one nearby sending the doves and starlings scattering.

Whenever you are abroad you don't have to go to the best nature reserves way out of town to see the best wildlife. It's often just a case of walking around a corner and opening your eyes.

Istanbul, Turkey

Never mix alcohol with birding – it just doesn't work. That was the first lesson I learnt at 5.30 a.m. on the first morning of my urban-birding trip to Istanbul. I was struggling to understand why my guide Murat had two separate sets of eyes that slowly drifted across his face. I must have looked like a confused chameleon as I tried to concentrate while he broke down the day's itinerary. The reason for my discombobulation? An unexpected binge that started in the wee hours of the night before with photographer Dean Eades, my companion, who had joined me on the trip to shoot some urban wildlife. It wasn't until 5 a.m. and six double vodkas and oranges later that I realised that I was trollied and that we were due to start our day thirty minutes later. A last-minute quick snooze had failed to sober me up; if anything, it had increased my general feeling of inebriation. Istanbul was one of those mystical places that I thought I'd never get to, yet there I was, standing in the pre-dawn late-September darkness feeling as sick as a dog and wanting to quietly die on the spot.

Murat was unaware of my dire predicament and thought that my subsequent regular naps/brief comas were due to tiredness. The first place on his list was one of Istanbul's major migration hotspots at Küçük Çamlica Hill to observe the legendary raptor movement over the Bosphorus. This hill was effectively a lightly wooded municipal park that afforded a superb elevated view across the city. I was about to potentially witness one of the greatest migration wonders in the world and I couldn't even see straight. I needed to sober up pretty quickly if I were to enjoy this spectacle that was expected to kick off once the thermals had started to rise later in the morning. So after barely managing to focus enough to see three out of perhaps fifteen or so migrant Red-breasted Flycatchers that were quickly flitting around in the trees, and unsteadily following four low-flying Hobbies overhead, I decided that it was time to properly slumber on a patch of ground that was frequented by a host of feral dogs and cats. Istanbul is overrun with these mostly friendly feral pets. The guys were laughing at me because the place where I chose to sleep was supposedly infested with fleas.

When I awoke scratching, over an hour later, I felt a thousand times better and, more importantly, functional and ready for some serious raptor watching. Moments later, I was watching the first of a wave of large birds passing overhead in the clear blue sky with a bunch of other birders from across Europe. Steppe Buzzards, Short-toed Eagles, and

Black and White Storks led the procession in waves across a broad front.

Küçük Çamlica Hill is a must for birders visiting Istanbul, as the raptor migration here can be sensational. One Turkish birder showed me some footage on his camera taken the week before of several thousand Lesser Spotted Eagles in the air together – a phenomenal sight. However, over the years the raptor counts over the Bosphorus have been steadily declining. In 1966, Richard Porter (co-author of *Birds of the Middle East*) and his friends counted over 250,000 soaring birds migrating over Istanbul from July to November. It was the first comprehensive count made of raptor and stork migration outside Hawk Mountain in the USA. In that year, he also found twenty pairs of nesting Black Kites in the city. Now there are none.

Later that afternoon we travelled north and stood on top of a nearby and relatively recently discovered watchpoint called Toygar Hill, which overlooks a valley next to the sea. Standing by the fire building on the brow of the hill we observed, often at closer quarters, well over a hundred low-flying Lesser Spotted Eagles, hoards of Steppe Buzzards, Short-toed Eagles, with a few Eurasian and Levant Sparrowhawks among them, a Goshawk and a Booted Eagle. Almost all of Europe's Lesser Spotted Eagles pass through Istanbul every year, as does the entire eastern European population of White Storks. Mind-blowing stuff.

We ended our first day by the coast of the Black Sea at Riva Stream, where plentiful Yellow-legged Gulls and Shags frolicked along the coastline, and flocks of Rock Buntings were in the seaside vegetation along with Sardinian Warblers and Stonechats. The reed beds by the stream itself housed the regular heron species, including some cracking Squacco Herons, a Marsh Harrier and passerines like Common Redstarts and Willow Warblers. There was also an interesting area of derelict land adjoining the reed bed that is supposedly good for Tawny Pipits, and on our visit it was covered with White Wagtails and a solitary Grey Wagtail, a local scarcity.

After my fateful first day I resorted to my more usual teetotal self and consequently was able to fully appreciate just how rich the city was as a birding venue. Perhaps my favourite urban birding areas were the Büyükçekmece and Küçükçekmece Lakes, both right in the middle of town. The former site was especially popular with gulls, which could be watched fairly closely from the historical and distinctly funky-looking Kanuni Sultan Suleyman Bridge. While sipping mint tea at the waterside cafe, we managed to scope five gull species, including several Slender-billed Gulls, plus Common and Whiskered Terns, Pygmy Cormorants, a Spoonbill and a roosting White Pelican – all in the middle of urbanity.

We also took a ferry across the Bosphorus Strait to the Golden Horn (the old city centre of Istanbul) to watch urban Alpine and Pallid Swifts swooping high over the heads of tourists, Laughing Doves and escaped Ring-necked and Alexandrine Parakeets. The ferry crossing yielded a tick for me in the shape of group of Yelkouan Shearwaters cutting above the waves through the flotilla of ships that litter the strait.

Istanbul is a captivating city, not least for its friendly people, interesting architecture and fascinating culture, especially given that half the city lies in Asia and the other half in Europe. It is also a very busy and crowded place, daunting unless you know where you're going or have a guide. I loved it.

Jerusalem, Israel

Israel is on a lot of birders' radars. Everybody's heard about the legendary birding to be had there, especially at famous sites like Eilat, in the south of the country, where quite spectacular falls of migrants occur. I was in Israel one autumn as a speaker at the international Hula Bird Festival, near the Syrian border, and the plan was for me to hang out for a week to enjoy the wintering 30,000 plus cranes as well as other great birds. Well, that was the programme. In the end, the lure of city lights was too strong and halfway through the week I decided to travel south to Jerusalem. I had left the Syrian Serins, Crimson-winged Finches and cruising Pallid Harriers in search of – well, who knows what.

So there I was, in a coach with a garrulous bunch of Jerusalem-based Jewish birders who had travelled to the Hula Valley the previous night to hear me talk. They had insisted that I join them for the journey home. On the way we stopped at the Sea of Galilee. Viewed from a distance it certainly looked biblical: a shimmering lake in the Jordan Rift Valley where once upon a time Jesus had famously walked on its waters. Now, I am not a particularly religious person but I did wonder if I would have an epiphany on the water's edge and throw my binoculars into the lake to become a monk. That did not happen. Instead, when I stood by the shore, or to be more precise on the elevated boardwalk, my jaw dropped. I was greeted not by a religious icon but instead by a very tacky touristy 'kiss-me-quick' scene with a sonic backdrop of loud, banging acid house. Fortunately, there were birds around to take my mind off the disappointment. A couple of Night Herons stood on the jetties patiently waiting for the next tour boat to arrive, while Armenian Gulls joined their Black-headed brethren on nearby rocks where there were lots of Pygmy Cormorants. A rather approachable Squacco Heron was stalking the rocks beneath the promenade that I was standing on, keeping a watchful eye on a nearby group of feral cats rummaging through a discarded falafel wrapper.

We reached Jerusalem by nightfall. The next morning I was taken to Jerusalem Bird Observatory in the heart of the city. Sitting on a hill sandwiched by the parliamentary buildings, a high court and an ornamental garden, it seemed an unlikely ringing station. Yet within this one-acre oasis, replete with artificial water features and the bush cover that you expect from an observatory, more than 200 species have been encountered and in excess of 7,000 birds are ringed annually. This is a staggering number of birds that are caught, including some great rarities like an Eyebrowed Thrush.

Founded in 1994, it is a community urban wildlife site that serves as a great educational resource for Israeli students and local children, who come along to watch and learn about birds and conservation. Their visitor centre was the first in Israel to be a 'living one' with a green roof and is festooned with seventy nestboxes for Great Tits, Hoopoes and sparrows. A school party arrived during my morning at the observatory and I sat excitedly with it as Spectacled Bulbuls and Palestine Sunbirds were extracted from the nets and processed. A few Robins and Song Thrushes were also caught; they are both winter visitors to the city and were the greyer continental versions of our own birds. Jerusalem was also experiencing a great Hawfinch winter. Like our Waxwings, Hawfinches occur each winter in varying numbers, and the signs were that this was going to be a bumper year. I easily saw around forty birds in the JBO as well as in the adjoining ornamental Rose Garden. I was also superthrilled to see them in the hand at close range. Interestingly, this winter the UK was also treated to a relative abundance of these cherry-stone-crunching behemoth finches.

Later, I popped into Gazelle Valley, a fifty-acre swathe of wilderness encircled by urbanity. It is home to a handful of the endangered Palestine race of Mountain Gazelles, whose number there had been decimated by local dogs. Thankfully, the area has been saved by the Society for the Protection of Nature in Israel and, over the next decade, the society plans to turn it into the country's first urban nature reserve, directly influenced by the London Wetland Centre. The western half of the city seemed to be quite lush with Mediterranean-type habitat, very different from the far more arid eastern side. That landscape was strewn with boulders and stunted trees. It seemed devoid of life. Once I got my eye in, however, with a lot of help and local knowledge from my guide, the barren vista began to give up its secrets and soon we were locating unobtrusive singing Long-billed Pipits, and more numerous Mourning and Finsch's Wheatears.

When I returned to England someone lambasted me on my blog for appearing to support the current political situation in Israel. My response was quite simple: I promote wildlife and not politics, and if politics dictated where I observed wildlife then I would never leave my home. Wildlife does not recognise political boundaries. Despite the difficulties in many countries across the world, there are still local conservationists fighting hard to save wildlife and they deserve our respect and support.

Nairobi, Kenya

The night I left Nairobi was a sad one. Yes, there wasn't a dry eye among us at the airport. I had spent the previous five days on an amazing dawn-to-dusk urban-birding roller coaster visiting haunts in and around this bustling capital. During that time I forged great friendships with some of the key members and guides of Nature Kenya who hosted me along with the Kenya Tourist Board. I also saw some fantastic birds (and mammals), and gained the sobriquet of 'Lucky Lindo' due to my perceived ability for making unexpected birds materialise seemingly at will.

From the moment I stepped off the plane from London I was swept up into the bosom of these dedicated conservationists. The main man was Mike Davidson, a retired expat banker who turned out to be one of the nicest people I've ever had the pleasure of meeting on my travels. Throughout the daylight hours I was continually flanked by him and up to four guides, who would identify practically every bird that I cared to glance at, plus call the many that I totally missed. I was equally overwhelmed by their immense knowledge and their humility. They made me feel as though I was a mate and not just a tourist who needed to be force-fed with lifers.

Nairobi is a very busy city filled to the brim with people. Once you look beyond the mass of humanity you will begin to see birds – stacks of them. Cattle Egrets congregate in roadside trees, Laughing Doves spirit from one spot to another, and omnipresent legions of Marabou Storks and Black Kites, sporting yellow bills, drift overhead. Some birders like to tick these kites as Yellow-billed Kites – a full species distinct from the nominate Black Kite. I'm with you guys. Nairobi is perhaps the best city in the world for urban birding with more than 600 species recorded. That is more than many countries' entire lists. In the days that I spent roaming in search of urban birds I saw over 300 species, of which about 100 were new. This city is not the sort of place for an urban birder to wander freely draped in the latest optics as there is the risk of crime. That said there are plenty of places to bird safe in the knowledge that the other people around will be watching birds, too, and not you. Many of the city parks are great places to check because of the sheer number of species to be found. It really can be quite bewildering.

One such inner-city site was Uhuru Gardens and Arboretum. It combined a mixture of boggy grassland with a miniature arboretum containing a high proportion of native trees. At first glance this small site didn't look as

though it was up to much. Indeed, buffering the area was a
busy main road, a major building site and Kibera, the
world's largest slum with an estimated one million residents.
Despite the very urban setting, it was jumping with
grassland and forest birds. I visited the park with a group of
local watchers led by famed birder and national legend
Fleur Ng'weno. French born and now in her seventies, she
has been leading her famous Wednesday bird walks in the
city every week for the past fourty-one years. Many of
Kenya's top birders started their ornithological apprenticeship under her tutelage. Fleur was certainly a charismatic
and unforgettable woman. She had more energy, enthusiasm
and passion for all things natural history than most people
a third of her age. What she didn't know about Nairobi's
birds was not worth worrying about.

One minute she could be describing the identifying
features of a Holub's Golden Weaver in a way that even my
mum would understand. Then, seconds later, she would stop
in her tracks to bring the assembled throng's attention to
some innocuous plant that had caught her eye. I stuck to her
like glue as we admired glamorous male White-winged
Widowbirds sitting on the tops of bushes, tracked a solo
Banded Martin as it patrolled the grassland and discovered
another avian surprise in the shape of an emerald-green
Klaas's Cuckoo – a bird that even she didn't expect us to find.
And the Harlequin Quail that we flushed from underfoot
was another eyebrow-raiser. They were only four of the
seventy or so species we encountered in our two-hour visit.

The remainder of my time was principally spent in Mike's
company along with my private posse of guides. Hanging
out with all those experts meant that my species list was
clocking up at an alarming rate. So much so that I barely
had time to write the name of a bird in my notebook before
the next lifer was announced. At Kinangop, a Nature Kenya
reserve, I had to be dragged kicking and screaming from the
Red-throated Wryneck that I was watching conveniently

sitting on a wall near the visitor centre, to help locate an ultra-rare Sharpe's Longclaw in a nearby field. In the short walk to that field we saw shedloads of birds and my notebook was burgeoning.

Kenya is a birders' paradise, but instead of racing through Nairobi to visit the hotspots like the Masai Mara, spend some time in the city. We spent a day in the Nairobi National Park on the outskirts of the city and registered nearly 200 species there alone. Now that's proper urban birding.

Port Aransas, Texas, USA

Texas was a bit of a surprise to me. As was the case with my original visions of Arizona, I imagined the Lone Star State to be a dustbowl, as portrayed by the likes of John Wayne cantering into a hail of gunfire on horseback. Cacti blotted the desert, from horizon to horizon. However, when I surveyed the scene while visiting the Rio Grande Valley, perhaps the USA's hottest birding hotspot, I realised how wrong I was. Situated in the south-east corner of the state and bordering Mexico, the valley is some 40 miles wide by 140 miles long, with more than 100 birding sites and a list of over 540 species. Astonishingly, I was looking at a landscape that was much greener and indeed lusher than I had ever imagined it to be.

It was November, and I had just spent a fabulous bird-filled week variously attending the world-famous Rio Grande Valley Bird Festival and going birding until I dropped. After the fair closed I took the advice of some local birders and went north to visit the locally raved about Port Aransas and its environs. A three-hour drive ensued. Although I was using a satnav, I had no clue as to where I was heading. However, the journey was not without action and drama, and I saw my first Sandhill Cranes in flocks that swept over the roof of my hired car in noisily gossiping groups.

So this is what Poole in Dorset would look if it was transplanted to coastal Texas, was among the thoughts that bubbled in my mind as I arrived in Port Aransas, thankful that the journey had ended. Have you ever heard of Port Aransas? It truth I had never heard of the place until this trip, but I was totally happy to be cruising its streets. Four hours drive from Houston and Austin, the nearest major cities, Port A (as it is locally known) is situated on the northern tip of the evocatively named Mustang Island, one of the longest barrier islands along the Texas coastline. Port A is the island's only established town and it really did remind me of Poole, having the same affluent feel as that south-coast English town. Here as well, you were never far from the sea. One of the first birds I saw, as I peered through the car windscreen looking for my hotel, was an Osprey. Over the ensuing days I literally saw masses of the birds. This majestic fish-eater winters along the Texan coast in large numbers, and I had seen hundreds on South Padre Island to the south the week before.

Port A is a small place – after spending a couple of days there I found the locals to be very hospitable, and the seafood was to die for. On the urban-birding front you could quite easily lose yourself exploring the plentiful bird-filled sites the town has to offer. Port A itself, along with the rest of Mustang Island, lies on a rich migration flyway. It consists of habitats that include lots of wetland, muddy inlets, coastal scrub and 18 miles of beach and dunes. This suite of habitats is perfect for the resident birds, along with the thousands of migrants that are also attracted here. Moreover, despite the area's popularity with birders and other recreational users, I found that you can often be alone with nature.

My favourite place for birding was definitely Port Aransas Nature Preserve. For me, this 1,200 acres of extensive tidal mudflats bordered by scrub is Port A's flagship birding venue. Part of the Great Texas Coastal Birding Trail, it is criss-crossed by elevated boardwalks and trails and has tower hides. The place is a haven for waders, egrets, wintering Ospreys and passerines like Eastern Meadowlarks – plus it plays host to the endangered Piping Plover.

There are two entrances to the site: one by Highway 361 and the other nearer Port Aransas proper. The latter leads into Charlie's Pasture, which was by far and away my favourite part of the preserve. I spent half a day watching clouds of Shoveler, American Wigeon, Pintail and Green-winged Teal launching themselves in the air to cut multiple shapes before dumping down on brackish pools. Although I was unsuccessful in finding any Wilson's or Piping Plovers, I was still treated to stunning views of Lesser Yellowlegs, Dunlins and Semipalmated Plovers. While squatting down on a boardwalk watching a Stilt Sandpiper feeding nearby, a Clapper Rail strolled out from under where I was stooped to feed quite unconcerned literally an arm's length away, much to my great surprise and delight. I even managed to find a hunting Short-eared Owl and a flyover Magnificent Frigatebird, both extreme scarcities here.

I found another urban-birding gem nearby: Joan and Scott Holt Paradise Pond. Situated behind a motel, this unlikely looking spot was decidedly very urban. It is a tiny, two-acre, wet woodland coastal marsh containing the only fresh water on the whole of Mustang Island. However, it is an amazing migrant trap, and no doubt its attraction is enhanced by the presence of fresh water. It was bordered by wooden fencing, on which I spotted a group of four cute Inca Doves cuddling up together. While creeping along the boardwalk I was joined by a couple of Scottish birders from Aberdeen. As it happened, their local RSPB group had only just received a talk from me the week before. They had managed to miss it so I whispered a vastly truncated version while we watched a lone juvenile American White Ibis feeding in a muddy channel. We also managed to identify a Pine Warbler that was busily flitting over our heads in a nearby waterside tree. It was particularly pleasing for me to see it as it was a new bird for the list.

Perhaps Port A's main claim to fame is the fact that it is the epicentre of the Whooping Crane Festival, which is attended by masses of locals and holidaymakers every February. Whooping Cranes spend the winter in the surrounding area. They are highly endangered, with just a few hundred birds remaining. I was lucky enough to have a personal tour of the fields outside the town for early-returning wintering birds. We found plenty of the far more common and smaller Sandhill Cranes, but it took three hours before we found a family party of Whooping Cranes. They were standing nervously in a field, heads up and alert. I slowly stood up on the roof of the pickup truck we were using to get a better look. My heart was racing while I watched this group of mystical birds as they occasionally uttered their trademark, bugle-like cries. They were huge, easily dwarfing the far more relaxed Sandhill Cranes that shared their field. Soon they took to the air. They were the perfect avian personification of grace, the black-and-white adults with their tan-necked juveniles appearing as if painted in oils against a sky-coloured canvas. This was a sight that I knew I was privileged to see. It is moments such as this that remind me of my love for birds.

São Paulo, Brazil

Passport. Check. Astrud Gilberto's *Girl from Ipanema* album downloaded onto my iPod. Check. Samba dance moves perfected. Check. And a field guide to the Brazilian birds packed. Double-check. I was totally ready for my first trip to Brazil. I was heading to São Paulo on a four-day visit to participate as keynote speaker at Avistar 2012, aka the Brazilian Birdfair. During the eleven-hour flight I had visions of walking along the streets of Brazil's largest city, skanking to the Latin beats blaring from everywhere while being bewildered by a procession of multicoloured birds cutting exotic shapes in the trees. I was preparing to have my head blown off, ornithologically speaking of course.

As the plane was touching down, I did what I always do when arriving at a new land: frantically scanned the airport grasslands looking for the customary first bird. To my dismay there was no obligatory corvid, hirundines or egret to greet me. Worryingly, I saw absolutely nothing and it was broad daylight. Far more fruitful was the trip from the airport to Parque Villa-Lobos, the venue of the fair. The skies were peppered with the dark, broad-winged shapes of soaring Black Vultures looking not so much for carcasses but for discarded garbage containing juicy morsels. Most of the green spaces and urban riverbanks we passed hosted Southern Lapwings – vociferous and flappy just like their northern counterparts.

It was Thursday and my talk was not due to be delivered until Friday, the opening day, so I took the opportunity to explore the parkland around the site of the fair. Parque Villa-Lobos was a largely manicured, city-encircled municipal park that seemed strangely silent and devoid of natural life. Well, in truth there were a few birds that I could not avoid seeing. Noisy Southern Lapwings patrolled every stretch of mown grassland, the ever-present Black Vultures rode the thermals above me and ubiquitous Blue-and-white Swallows, which looked like House Martins minus the white rumps, swished after unseen insects. Hunting on the ground and resembling robust female Blackbirds with rich ruddy underparts were Rufous-bellied Thrushes alongside very confiding Rufous Horneros. The latter species was a curious sandy-coloured bird that had a Starling shape and walk, a Nightingale rufous tail, and flew like a woodpecker. But despite these great birds I still felt cheated. Where were the masses of monkeys, general junglely noises and the technicolor, multi-species tanager flocks that I had been promised?

I soon learnt from the São Paulo birders that as we were rapidly approaching winter many of the birds I desired had already hot-tailed it out of town to more northerly climes. I had also failed to appreciate that neotropical birding was far harder than I imagined. Some of the forest birds have a remarkable ability to sit still and melt into the sparsest of cover. So with the birds either sitting tight or hundreds of miles away, surely my luck was out?

Unperturbed, a couple of days later I joined a thirty-strong group of urban birders on a walk in Parque Ibirapuera right in the heart of the city. This urban park was comparable to New York's Central Park and London's Hyde Park – overrun with joggers, cyclists and dog walkers. But despite the masses of people there were loads of birds occupying some fairly decent habitat that included a large central lake, woodland and gardens. The lake held small multitudes of Neotropical Cormorants consorting with White-faced Whistling Ducks and the very same Moorhens that we have in the UK. On the shoreline stood a congregation of large and ugly-faced Black Vultures seemingly indulging in some kind of committee meeting. Stalking the water's edge were Snowy and Great Egrets, and I also noticed a Striated Heron that had secreted itself deep within a waterside shrub.

There was plenty to see on dry land too. Plain Parakeets were very approachable, precariously balanced as they pecked at fruit in the same trees that sky-blue Sayaca Tanagers foraged. My group got really excited when a Rufous-browed Peppershrike was discovered singing deep in a woodland canopy. I had no clue what this discovery meant, as I had not even heard of one before. My favourite moment was when a pair of Masked Water Tyrants materialised on top of a bush near the lake. They were like grey-and-white Wheatears with a black bandit mask and a vaguely similar tail pattern. The resemblance was uncanny, yet they both belong to totally unrelated families.

I was not able to explore São Paulo's urban expanses for birding sites as completely as I wanted. This was partially due to my commitment to Avistar 2012 and the fact that the city is potentially not the safest place to be wandering around with optics. That said, if you ever find yourself there you could also do no worse than to check out the Zoological Gardens and Botanical Gardens in the south of the city. Both are on the beaten track and are faithful refuges for urban birders.

Although São Paulo was not the most picturesque of cities, the birders I met were warm, helpful and very welcoming. I now have about thirty new Facebook mates and enough places to stay in the city and along the Atlantic Forest coast to last a six-month sabbatical.

Although I never got to samba or to encounter the fabled flocks of neotropical wonders, I had a marvellous experience nonetheless. It would have to be a summer visit next time, methinks.

Taipei, Taiwan

'Whatever you do, don't mention the number four. It means death in this country.' I gulped when my guide uttered those words as I focused on four Pacific Swallows sitting on a Taipei telegraph wire. I was in the capital of Taiwan in the company of my guide Kuen-Dar Chiang (also known as Ata, for short), scouring the streets and parks to discover the best of the birdlife this Asian metropolis has to offer. Taipei certainly was a big city, home to well over eight million people with millions more in the outlying satellite settlements that encircle it. To be honest, I didn't really know much about Taiwan prior to my visit. I knew that it was an island next to mainland China and that the country's name was often prefixed with the words 'Made in… ' – in other words, I was pretty ignorant.

That morning I got up at 4 a.m. feeling pretty caned after sleeping for just three hours after being on a fourteen-hour flight. Leaving the hotel, I stepped out for a walk through the waking urban streets. The first bird I saw was a Tree Sparrow. It soon became apparent that as in other Asian cities this strictly rural sparrow in Britain was a complete urbanite here. Dainty Spotted Doves with their intricately freckled neck patches were also abundant, along with lots of Barn Swallows – the same birds that we have at home albeit with paler underparts. They were swooping around some of the busiest streets, with pairs nesting openly in the eves on the inner-city houses and shops.

I then happened across a municipal park in the centre of town filled with Taiwanese pensioners who were variously exercising. Some were all wearing the same kit and were doing moves communally, waving their arms roughly in unison to the vocal commands of an equally ancient leader. Others were more solitary, trotting or walking around the park while self-beating their arms or backs. I started scanning a small marshy pond in the centre of the park, desperately trying to ignore the group of gyrating grannies next to me, and soon discovered a remarkable-looking Malayan Night Heron. Much more tan coloured than the Night Herons that we get in Europe, this bird seemed almost reptilian as it stood in the open, making itself invisible under a nearby tree.

Later that morning Ata took me to Daan Park, an open space of a similar design to the one I had visited earlier. It, too, had its own population of friendly stretching OAPs and a whole range of birds. We found a couple of Crested

Goshawks that apparently nest in the sparse tree cover, and enjoyed great views of local residents like Chinese Bulbuls and a small Black-crowned Night Heron colony on a shrubby island in the small lake. For me by far the best birds were the Oriental Turtle Doves, a species that I had lusted after for years. It was so great to watch lots of them in their natural environment. I was sweating profusely because of the heavy humidity but that made a change from queuing up, freezing outside someone's back garden in Oxford during the middle of winter in the hope of glimpsing a stray.

It was in the botanical gardens where I noticed another Taipei peculiarity. Occasionally, we would come across a crowd of men patiently waiting by some trees or at a clearing, photographic lenses poised as if staking out some major rarity. In reality, these guys were not birders but photographers after *the* shot of any photogenic bird in the vicinity. In this case they were after a snap of one of Taiwan's many escaped exotics – an Oriental Magpie-robin, a handsome pied chat but, nonetheless, an escapee. Bizarrely, they fully ignored other, less gaudy native species and they didn't have any interest in watching behaviour. I did find a posse of them 'papping' a scantily clad woman they had just discovered sitting on a bench by some bushes – or so it seemed. They were purely trophy hunters. If only the fellas in the Mediterranean would take the lead from these guys and put down their guns and pick up a 600-mm lens instead.

My favourite Taipei venue had to be Guandu Nature Reserve, run by the Wild Bird Society of Taipei. It consisted of fifty-seven acres of well-thought-out paths, hides and ponds surrounded by an additional area of riverine mudflats, rice fields and marshland. When we arrived an almighty electrical storm was brewing. Dark, ultra-menacing clouds were creeping ominously towards us over the mountains that skirt the city. The marshes are a great place for waterbirds. A Common Kingfisher whizzed past as we watched masses of Little and Eastern Cattle Egrets with a few feral Sacred

Ibises flying to their roosts. Dashing Pacific Swallows joined the swirling Barn Swallows. They shared the classic swallow colour scheme but had almost non-existent streamers with broader and slightly longer wings. The wader department was well represented by a few Red-necked Stints, a couple of Marsh Sandpipers and a passing Common Sandpiper that added a touch of familiarity.

Taipei was truly an urban-birding gem. My experience was made all the more special by Ata's expertise. He had a lovely chubby little face and a big smile that never left it the entire time that I was with him. He kept smiling even when I lied and told him that I had five Pacific Swallows on that Taipei telegraph wire.

Tucson, Arizona, USA

It was a dark, still and decidedly cold night when I rode into the dusty, arid outskirts of Tucson, Arizona. The previous week I had been enjoying urban Ravens and whizzing Anna's Hummingbirds in Los Angeles, but the time had come for me to migrate to the next stop in my impromptu tour of the US. I looked up. The sky above was pebble-dashed with innumerable stars. It was overwhelming. Their sheer number and density left me feeling crushed into insignificance. On the roadside ahead a lone moth-eaten Coyote foraged, pausing to size me up as I passed by. I was not riding into town on one of those piebald steeds that the Apache Indians popularly used in films, but in the back of a Toyota Prius. I was being driven by my host Sharon Arkin, an eccentric older lady who I had met six months previously in a similarly arid terrain in Israel.

I first ran into her while with a group of international birders watching several Marsh Harriers and Merlins coming in to roost in the Hula Valley, northern Israel. All was good until one of our number dropped a pair of glasses down a rocky ravine. Instinctively, I jumped into the crevice to retrieve them with all the grace of an overweight Chamois. The owner of said spectacles, Sharon, and I instantly became mates and she invited me to stay at her Tucson B&B, if ever I were in town. So, I took her up on her invitation at the last minute when I came to the US on a coast-to-coast round of speaking engagements. Sharon had planned a series of talks and walks for my five-day stay, so after my late-night arrival I excitedly arose at dawn the next morning to go and explore the immediate area around her house.

Sitting on many of the telegraph wires were pairs of 'Mo Does', or Mourning Doves to you and I. A ubiquitous species across the US, this medium-sized, long-tailed pigeon always reminds me of our Collared Dove when it indulges in its gliding display flights. It didn't take long for me to register my first new species: a Gila Woodpecker on the trunk of a Saguaro Cactus, the classic armed cactus that you see in Westerns. This south-eastern US desert woodpecker was quite distinctive: looking a little like one of the several flicker species found in the US, it had strongly black-and-white barred wings that when it flew had me thinking that I was watching a Hoopoe. The Gila (pronounced 'healer') turned out to be a fairly commonly seen bird around the edges of the city. I was pretty trigger-happy with my camera trying to get a close-up shot of one. I managed to get a bird on the end of a telegraph pole. It wasn't until I scanned through the images later that I realised that I had photographed a Gilded Flicker, a superficially similar and far scarcer species that turned the heads of the Tucson birders when I told them about it.

One of the outings Sharon organised for me was a group trip to Sabino Canyon. Although a popular spot, it was an unforgiving landscape consisting of the default Saguaro

Cacti and plenty of extremely thorny scrub interspersed with a criss-cross of paths. Lifers abounded. Black-throated Sparrows shared scrubby bushes with the occasional exquisitely named Pyrrhuloxia. No, it wasn't a disease but the less red southern relative of the more well-known Northern Cardinal. While birding around some riverine scrub, I spotted yet another lifer.

'Bridled Tit!' I exclaimed as I pointed to a bird not dissimilar to a Crested Tit deep within a leafless bush. My comment was met with disapproving looks. Had I got my identification wrong?

'We don't say that around here. It's either a titmouse or a chickadee,' came a voice from the crowd. To add insult to injury I then told the story of the Bearded Tits that showed up in a tiny reed bed in Hyde Park, London. They were far from impressed when I told them that I photographed them, placed the images on my blog and entitled it: 'I saw a pair of false tits in Hyde Park.'

By far my favourite birding spot in the city was Sweetwater Wetlands – a sewage-treatment works. It is the city's hotspot for rarities, no doubt because it is the only reasonably sized area with standing water for miles. Wandering around with the local birding group I was soon surprised by the sheer number of waterfowl. Hundreds of Shovelers, Pintails and American Wigeon shared the flooded basins with American Coots, and around the muddy edges were Least Sandpipers, Black-necked Stilts and a few Killdeer. In the reedy fringes of some of the watercourses were calling Soras that crept unseen, more obvious Common Gallinules (recently split from our familiar Moorhen) and singing Marsh Wrens. The best bird for me was an overwintering Solitary Sandpiper that had chosen to frequent a small channel. It was unusual here as they are normally scarce passage migrants. Sweetwater Wetlands is certainly a local patch to die for.

Speaking of dying, by this point of my trip – the penultimate day – I had contracted a vicious virus and I was

left feeling pretty slaughtered. Regardless, at the end of my
visit to Tucson I headed to the airport to fly to Philadelphia
to hit my next destination, the fabled Cape May. But the
way I was feeling, I wasn't sure if I would survive the flight.

West Hollywood, USA

Los Angeles is the City of Angels. Perhaps, surprisingly, it is
also a city of birds. It's the second largest city in the US and
is a massive, sprawling place with a population in excess of
fourteen million. It's a true 'megacity' in global terms. The
classic image that most people have of LA is great weather,
big flash cars and beautiful, physically augmented people,
most of whom are wannabe actors and Hollywood stars.
But there is also plenty to excite the urban birder. LA is
made up of a collection of different cities, including West
Hollywood – a city whose salubrious streets and surrounding
areas, such as Beverly Hills and coastal Santa Monica, I have
been regularly visiting for more than ten years.

It's true to say that nobody walks very far in LA. Practically
everyone drives everywhere. If you were to take a stroll in
West Hollywood you would notice that even the feral pigeons
seem far more lethargic than our birds in Britain as they loaf
on top of the massive billboards. You will also quickly begin
to see the plentiful American Crows and Ravens swooping
over the streets. The Ravens in California seem smaller than
the birds that we are used to and there is talk among the
scientific community of splitting them as separate species. You
will no doubt see Western Gulls flying around over the city.
These birds are like broad-bodied Lesser Black-backs with
shorter wings. As well as the common Mourning Doves you

may be lucky and catch sight of the much shyer Band-tailed Pigeon that has the jizz of a slim Woodpigeon.

West Hollywood is bordered to the north by the Hollywood Hills. This area serves as a sanctuary for many of the film world's glitterati and is strewn with big houses, many of which look pretty bizarre. It is also the location for one of my LA birding stamping grounds, Franklin Canyon Park. It's essentially a fairly large, wooded valley encircled by a road with a small, natural-looking reservoir at the southern end. One summer's day I was birding from the road, watching a singing Song Sparrow, when I was tooted by someone driving past in an open-topped Bentley. The rotund, bubbly smiling face behind the wheel belonged to none other than chat-show host Jay Leno.

During any season you should easily see Red-tailed Hawks circling overhead and the occasional Cooper's Hawk, which is a large *Accipiter* midway in size between our familiar Sparrowhawk and a Goshawk. Western Scrub Jays are common in the wooded areas, along with roving parties of Bush Tits that remind me of tiny brown Long-tailed Tits. Pied-billed Grebes, Wood Ducks, obligatory Mallards, and both Green and Great Blue Herons are fairly easily found on the reservoir, while in the summer chattering parties of Barn and Northern Rough-winged Swallows congregate on the nearby telegraph wires. In winter it's an excellent spot for flocks of Cedar Waxwings.

Another area to check out is Kenneth Hahn State Recreation Area on the Baldwin Hills. Much loved by a platoon of wealthy-looking joggers, an early-morning summer visit will reward you with Anna's and Allen's Hummingbirds, Lesser Goldfinches, and Say's and Black Phoebes. Check the grassland areas in the autumn and winter for roving American Robins, and on a couple of occasions I have come across the regionally scarce Western Kingbird and the nationally rare Loggerhead Shrike – a bird that looks like a smallish Great Grey Shrike.

Speaking of rarities, if you want to stand a better chance of finding something interesting, especially during the autumn migration, there are several places to discover. Perhaps the most well known among the LA birders is Harbor Park, a coastal park with a relatively large lake surrounded by reed beds and municipal parkland. Aside from egrets and several heron species, including Night Herons, you should also come across some interesting duck species, such as Cinnamon Teal, and skulking Marsh Wrens in the reeds. There are relatively few trees in the park area but if you visit on a good autumn day the scattered trees are literally alive with migrants, particularly American wood warblers, and occasionally they can include east-coast stragglers like Black-and-white Warblers.

My favourite LA birding location, however, is the Ballona Wetlands (pronounced bi-yowna) on the coast close to LAX (the city's international airport). I always make a beeline for this 1,087-acre ecosystem whenever I'm in the city, as it's only a short thirty-minute drive from West Hollywood. The birdlife here can sometimes be incredible and it still amazes me how close you can get to some of the birds. On the saltwater and freshwater marshes, depending on the season, you can sometimes get fantastic views of the visiting waders. I've seen both Long and Short-billed Dowitchers, and Western and Least Sandpipers at ridiculously close range. You can also see resting Ring-billed Gulls sitting side by side with Brown Pelicans.

The adjacent beach, which can sometimes resemble a *Baywatch* set if you arrive during the day, is usually coated with mixed gull and tern flocks, with Royal and Elegant Terns rubbing shoulders with Western, California, American Herring and Heermann's Gulls. During the winter Glaucous-winged and Bonaparte's Gulls arrive, and on the sea there are many Western Grebes and Surf Scoters.

So as you can see, there's more to West Hollywood than the glamorous veneer that first meets the eye. Almost any semi-wild area will produce a gem that you were not expecting. In LA it certainly still is the Wild Wild West.

Yucatan, Mexico

I must have been driving for nearly three hours at five mph along the most treacherous road in the world, continually being slapped around in my little tin box of a car. There were all the classic bits of drama to add to the mix, too; I was dog tired after driving all day, and I was hungry and very thirsty, with nothing to sustain me in the car, which, incidentally, was running dangerously low on petrol. I was in the wilds in the pitch black of night and a little scared, if I were to be honest. I was not even travelling on a proper road. It wasn't a track either, rather a dusty path just wide enough for my small tin car and indiscriminately peppered with massive craters. It was March 2004 and I was heading to Punta Allen on the Boca Paila Peninsula, Yucatan, Mexico, in a white Fiat Punto. Alone.

It was the first time I had ever travelled anywhere far flung on my own and, though I was not afraid of making the trip in the first place, I was genuinely worried for my safety. To my left in the murk was the sea and on the other side of me was thick dark jungle. I didn't know where or what I was heading to and I could not even turn around because the road was too narrow. When I wound the windows down strange, shrill noises and guttural shrieks uttered by invisible beings filled the car – the faceless denizens of the dark, echoing from the jungle. I kept on having visions of being ambushed by lawless bandits, shot and left to rot, for my remains to be eventually found fossilised in my Fiat Punto a hundred years hence. 'Make sure you don't drive this vehicle off-road over potholes,' said the hire-car man in broken English back in Tulum as he handed me the keys. Why didn't I listen to him? At the time my only concerns were why white? Why couldn't he have given me a cooler and darker car that wouldn't show up the dirt so much?

When I started my fateful journey the evening sun was about to set. It was late to be travelling but someone I had met back in England had told me that Punta Allen was just down the road from Tulum. Down the road is just down the road, I reasoned. Apparently Punta Allen was an interesting, friendly, hippy commune that lived off the land and generated its own electricity. More importantly, the commune was situated on a bird-rich peninsula. I couldn't resist. My plan was to make the trip and find some accommodation at the other end.

I had landed in Cancun a week before, having originally flown out to meet an ex-girlfriend who was travelling around South America. At the last minute before I was due to board the plane from Heathrow she reneged on our arrangement, believing that I wasn't coming, and travelled on. I was bitterly disappointed but I was not going to waste a £700 air ticket, so I decided to head out despite having nowhere to stay. Cancun was nasty. I developed an instant dislike for it as soon as I arrived under cover of darkness. It was noisy and seemed like a complete and utter tourist trap, so after examining a map at the airport I caught a bus to Pueblo Morelos, a small coastal town some eighteen miles to the south. I booked into the first hotel I came across and settled in. I couldn't sleep as I was too excited by the prospect of birding on a totally new continent, so I arose at 4 a.m. Stepping out into the darkness for a walk around the immediate neighourhood I began to hear weird noises emanating from the nearby jungle, while almost underfoot bucketloads of small, pretty frogs chirruped. A large bat with a wingspan comparable to that of a Woodpigeon began to circle me. Unnervingly, it silently flew around me in ever-decreasing circles, sometimes illuminated by the streetlights before slipping into the darkness. I had visions of being bitten by a rabid bat and ending my days in a frothing fury chasing monkeys around the jungle. It was time to retreat back to the hotel.

On the way to my room I noticed that the open-plan restaurant area was right on the beach. So I grabbed my bins and sat on a hammock to await the sunrise. The first birds I saw were Great-tailed Grackles displaying on the nearby rooftops, the males looking like miniature black pheasants in the pre-dawn light. They would throw their heads skywards to emit an amazing array of mechanical-sounding calls. In the ensuing days this species would prove to be the most ubiquitous bird in Yucatan. As I gazed seawards I could make out the unmistakable shapes of diving Brown Pelicans fishing out over the waves. How could a big bird with such a large bill dive so effortlessly?

Already the tensions of London life were beginning to drain from me and by the time the sun had risen and I had breakfasted, my watch had been banished to the hotel room. OK, I still had my mobile phone with me. Old habits die hard. I spent the next few days travelling at a slow pace and birding all day long. Truckloads of birds went by unrecognised as I struggled to identify the confusingly similar-looking empid flycatchers, the baffling decurved-billed woodcreepers and other tricky South American families. I quickly devised a system whereby I made copious notes on each weird new bird that I saw. Salient identification features, descriptions of behavioural peculiarities and some pretty crude pen drawings made it into my notebook. Every night I would try to match my observations with the species detailed in my guidebook, with varying success.

I hired the car in Tulum, a famous tourist centre much beloved by visitors wanting to see the Mayan ruins in the town. It is apparently the third most visited archaeological site in the whole of Mexico. As much as I was interested in seeing Tulum's ruins my plan was to drive to Lake Coba, some fifty miles to the north-east, to try to track down the locally scarce Spotted Rail. Although fairly widely distributed from Mexico through to Central America, it is only regularly seen in a few places and certainly at Lake Coba, it hasn't been recorded since the Eighties. The odds were stacked against me but I do like a challenge. A stroll around the lake at first light triumphantly resulted in a fantastic sighting of a Spotted Rail and at a ridiculously close range. Of course, I had left my camera at the hotel.

I had a brief brush with some ruins on the beautiful Isla Cozumel a short ferry ride from another coastal tourist hell-hole, Playa del Carmen. On the ferry I had the dubious pleasure of sitting with some Louisianans who claimed to be Britney Spear's relatives. They took great pleasure in denouncing her evil ways and unsavoury lifestyle. I could not let them talk about our Britney in that way, so I tried to change the subject by pointing out how beautiful the sea was. Deep, clear and blue. Now totally distracted I turned their attention to a group of flying fish that had chosen to momentarily swim alongside the boat. They launched themselves airborne, seemingly flapping their elongated pectoral fins in brief flight before gracefully falling back into the sea. I had never seen flying fish before; they looked so much like birds. My new Louisianan friends were then to spend the rest of the journey talking about nature. That was more like it.

On Cozumel I hired a clapped-out Volkswagen Beetle for the day to do a loop around the island. I picked up a couple of the island's endemics, including the absolutely gorgeous

Cozumel Emerald, a distinctive hummingbird with an emerald-green body, long black wings and a black, deeply forked tail. When I got to the ruins I had every intention of slowly walking around examining the architecture and admiring the workmanship. Instead, I was mesmerised by a series of large, colourful butterflies fluttering past, and a range of bugs and lizards scuttling on the ground. Most amazing, though, was the Turquoise-browed Motmot that I discovered catching bees in a lightly forested area around the ruins. It was an incredible bird to behold – a heady mix of turquoise, russet and green with a blue tail adorned with long, bare tail shafts tipped by gorgeous rackets. It looked like an overgrown Bee-eater. Unmistakable, you would have thought, but the moment it landed it just melted away and became invisible, even fairly close to me at head height in the most sparsely vegetated tree. Like a lot of the jungle birds that I encountered, it seemed to share the habit of sitting stock-still for ages, quite unlike the furtive searching that I had been expecting.

They were nice memories but they did little to quell my desperation as I battled along the never-ending potholed road giving the chassis of the car an unholy battering. I noticed a creature in the road ahead illuminated by my headlamps. It was just standing there glaring at me, so I stopped the car to try to identify it. The ringed tail confirmed that it was a Coatimundi, a slender, lemur-like member of the raccoon family. It seemed to grin at me as it slunk off into the darkness. Was it going to be feasting on my expired body later on in the night? I pressed on at my ultra-slow speed. The most tiring thing was the continual swerving around craters to avoid doing my undercarriage any further damage.

It was midnight and I had been driving at a snail's pace for the past five hours. I had a few false dawns when the track got a bit wider, signalling that it was about to become a road. Could there be the possibility of civilisation around the corner? No, I was to be thwarted again when the next bend reverted back into the hellish road. I was giving up hope. Was I really going to end up in Argentina? Just as suddenly as my hell started it was over. I rounded a corner and found myself driving in a small, darkened town – Punta Allen. Since it was after midnight I guessed that the generator had done its work for the day. I also thought that I probably would not find a bed for the night, so I drove around looking for a spot to park the car and sleep. Then, I noticed a restaurant illuminated by what seemed like candlelight. I parked outside and approached the door. I could see through the window a bunch of burly guys chatting with a young woman. I pushed the door open despite the closed sign. The classic scene played out in which all the conversation and laughing ceased and all eyes were on me.

'Good evening, sorry to bother you guys, but do you know anywhere that I could get a bed tonight?' I enquired.

'No,' grunted one of the guys. He was American, and on closer inspection they all looked like rednecks defined in my mind by their lumberjack shirts and thickset builds. I offered an apology and turned to walk out of the door. The woman ran after me.

'You can stay in the cabana out back for ten bucks.'

What's a cabana? It didn't matter. I was so tired I would have paid $100 to sleep under a table in a dank toilet. I thanked her, grabbed my rucksack from the car and followed her to the back of the restaurant. In the pitch black she led me, to what looked like, a series of large witches' hats. We entered the last one. The interior was obviously cone shaped with a bed draped with a mosquito net in the middle. After she left I wearily did a cursory search of the bed for any unwanted partners and fell asleep before I could assess the possibility of being murdered during the night. I woke up at dawn, alive. Shortly after I was out looking for birds. By lunchtime I had hit a brick wall. I was exhausted and disappointed. Punta Allen was not as amazing as I thought it might have been. As it was a thin strip of land jutting out into the Caribbean Sea the beaches were polluted with loads of litter and other debris brought in by the tide. The birding was equally disappointing, though I did get a couple of new ticks, including a Yellow-billed Cacique, a type of grackle, and perhaps one of the most stylish birds of all, a Canivet's Emerald. It was the most beautiful hummer I had ever seen – brilliant iridescent green with a deeply forked dark tail and a black-tipped red bill. At one point I had it hovering right next to me. It was a truly incredible sight and easily the highlight of my Punta Allen experience.

I made the journey back down Hell Avenue after lunch, despite being invited to stay longer for some more birding by an American couple who took a shine to me. The way back always seems shorter, especially when you know how long the journey is and, crucially, you are doing it during daylight. And with a full tank I upped the speed. The American couple had told me that I had driven around forty miles last night and had burst out laughing when they saw that I did it using a by now tan-coloured Punto.

When I arrived at Tulum my first port of call was the local car wash. After restoring the car to its rightful gleaming white glory I returned it, clanking somewhat, to the hire company. I think that they were not used to people returning their cars clean, so they neglected to check the car over. Result. Rucksack on back, I caught a bus and travelled back to Pueblo Morelos, my starting point, for the last couple of days of my trip, where I managed to lose my mobile phone up a tree hide in the local botanical gardens. No doubt there is a Spider Monkey out there having a great time ringing up all its relatives. As I sat on the plane heading home, I resolved that I must travel more often, but not just to the middle of nowhere.

Epilogue

There is not a day when I don't marvel at the nature that surrounds me in my urban environment. Birds are my life, my love, my sanctuary and my therapy. Without them I don't think I could survive. For me, every day is an amazing day. I look up and may see a Cormorant purposefully flap overhead or hear a Great Tit singing in the spring, its repetitive song sounding like one of those old-fashioned clock-radio alarms. Even a Woodpigeon lazily half-waddling, half-flapping out of my path, as I accidently tread on the crumbs that it was previously pecking, fills me with joy. I would be so sad if those everyday occurrences were to cease.

This source of joy and peace can be accessed by anyone anywhere in the world. The best part of it is that it is absolutely free. What's more, it can be encouraged to flourish and prosper, which in turn makes it even more readily available for all to see.

Any one of us can fall in love. Every one of us should fall in love. Not just once but many times a day. So go ahead and fall in love. The birds will love you back unconditionally and will continue to fascinate you until you draw your last breath.

Acknowledgements

This book and, indeed, the adventures that I have had in my ongoing urban birding quest would not have been possible without the belief, support and editorial platform provided by *Bird Watching* magazine. I am especially indebted to David Cromack, Kevin Wilmott, Sheena Harvey and Matthew Merritt – editors past and present.

I'd also like to thank the host of people who helped me in so many ways to tell my stories, forgive me if I've forgotten anyone: Margaret Adamson, Peter Alfrey, Dave Allen, Dr Sharon Arkin, Martha Argel, Steve Ashton, Mark Atkinson, Amir Balaban, Mark & Gill Baker, Xana Batista, Cynthia Bendickson, Stephen Boddington, Murat Bozdogan, Emily Broad, Paul Brook, Chris Brown, Nick Brown, Roger Brown, Guto Carvalho, Sharon Cavanagh, John Charman, Kuen-Dar Chiang, Mr Chiang, Toby Collett, Jim Coyle, Megan & Mike Crewe, Richard Crossley, Mike Davidson, Abiy Dagne, Ed Drewitt, Kim Dixon, Sarah Doble, Malcolm Duck, Andrew Easton, Collin Flapper, Richard Fray, Dave Gandy, Roland Gauvian, Iain Gibson, Gerard Gorman, John Hague, Penny Hayhurst, Sheri Henneberger Peter Herkenrath, Martin Hierck, Sam & Tina Hobson, Claire Holder, Ben Hurley, João Jara, Les Johnson, Steve Jones, Martin Kelsey, Jip Louwe Kooijmans, Olive May Lindo, Jim Lawrence, Neil Lawton, Anthony McGeehan, Sara McMahon, Frederic Malher, Luke Massey, Jeff Mears, Jonathan Meyrav, Gerby Michielsen, Nick Moyes, Pete Naylor, Fleur Ng'weno, Simon Nichols, Marc Outten, Zoo Park, Dr Gidon Perlman, Shaun Radcliffe, Mark Reeder, Alastair Riley, Avner Rinot, Phil Round, Dragan Simic, Rick Simpson, David Stubbs, Hrafn Svavarsson, Hannu Tammelin, Claire Thomas, Alan Tilmouth, Simon Tonkin, Barry Trevis, Sedley Underwood, Peter Wairasho, Mike Weedon, Kevin Wilson, Tony Whitehead, Darren Woodhead.

Index